The Natural Dyer's Almanac

The Natural Dyer's Almanac

A year of natural dyeing

Caroline Bawn

YOUCAXTON
PUBLICATIONS

Copyright © Caroline Bawn 2022

The Author asserts the moral right to
be identified as the author of this work.

ISBN 978-1-914424-65-6
Published by YouCaxton Publications 2022

All rights reserved. No part of this publication may be reproduced, stored in a retrieval system, or transmitted in any form or by any means, electronic, mechanical, photocopying, recording or otherwise, without the prior permission of the author.

This book is sold subject to the condition that it shall not, by way of trade or otherwise, be lent, resold, hired out or otherwise circulated without the author's prior consent in any form of binding or cover other than that in which it is published and without a similar condition including this condition being imposed on the subsequent purchaser.

YouCaxton Publications
www.youcaxton.co.uk

ABOUT THIS BOOK....

The aim of this book is to enable you to feel confident to have a go at dyeing some yarn or fibre with natural dyes that you might find at home, in the hedgerows or in the garden, or that you might grow yourself. I'm based in Cornwall, in the UK, so my plant-growing and plant-foraging advice is for the UK and perhaps some of the northern hemisphere.

The book is set out in monthly sections and each month has information about the dye plants themselves, the way to grow them (if appropriate) and how to use them to create colour. The first 4 months of this year use colours from your larder or store cupboard, the second 4 months use foraged dye stuff, and the last 4 months of the year use dye stuffs you might have grown or can buy. Each month a new technique is introduced to help you build your skills and knowledge throughout the year. At the end of some months there are recipes, just for fun, and through the book are small, simple patterns to knit and crochet.

I use the terms 'yarn' and 'fibre' interchangeably. Both can be successfully dyed naturally and I hope this book will appeal to spinners, felters and textile artists and well as knitters and crocheters. Most of the dyes in this book have been used on wool-based yarns unless otherwise stated. I also use the terms dye pot and dye pan interchangeably.

THE BOOK EXPLORES:

Mordants – the way to help colour fix to yarn and stop it fading quickly or washing out, and which types of fibre need which mordant.

Substantive and fugitive dyes – some dyes will give long-lasting and reliable results, and some will wash out and fade.

How to prepare dye – there are several ways to prepare dye, depending on the plant source

Adjusting the dye colours – using acid (vinegar) or alkali (bicarbonate of soda) to change the colours of some natural dyes, and some metal solutions can also be effective.

Different ways to use dye colours – learn how to dye with a traditional dye pot and also in a large kilner-type jar.

Dye plants you can forage – the hedgerows have dye plants available throughout the year. They can be used to create colours that have been used for centuries.

Dye plants you can grow – a sunny spot on a patio or balcony is enough to have a go at growing a dye plant

Contents

About this book....	v
The Natural Yarn Dyer's Almanac	ix
Guidelines of responsible natural dye foraging and growing	xi
Glossary of Terms	xvii
Planting and growing guide	xxi

January — 1
- January in the dye garden — 2
- Choosing and buying seeds — 2
- What Next? — 2
- The Colours — 3
- Dyeing with Turmeric — 4
- To Dye your Yarn with Turmeric: — 4
- This Month's Pattern — 6
- Ariel Shawl — 6
- About Turmeric — 7
- Sfouf Cake — 9

February — 11
- February in the dye garden — 12
- Pots — 12
- Mordants — 12
- Alum — 12
- How to Mordant your fibres — 13
- Cold Mordanting — 13
- Hot Mordanting — 13
- Avocado dye experiment — 15
- Black bean chilli — 18
- Pattern: a simple stash-busting Cowl — 19

March — 21
- In the Dye Garden — 21
- I have planted — 21
- More Mordants — 22
- Next comes the maths — 22
- Ginger biscuits — 23
- Red cabbage Dye trial — 24
- Here's what I did — 24
- Using pH to change the dye colours of Madder — 25
- This Month's Pattern — 26
- March Mittens. — 26
- Nettle socks lace chart — 28

April — 29
- In the Dye Garden — 30
- Mordanting Experiments — 30
- The cotton scour/mordant experiment goes like this — 31
- A bit about Rooibos — 32
- Dyeing with Rooibos — 32
- Here's the Delta Shawl Pattern — 33
- Rooibos Shortbread — 35

May — 37
- In the Dye Garden — 37
- Gorse — 38
- Dyeing with Gorse Flowers — 38
- To prepare the dye — 39
- About Dye concentrations — 39
- Gorse Flower Cordial Recipe — 40
- Gorse Blossom Cardigan — 41

June — 45
- In the Dye Garden — 45
- Exhaust Dyeing — 46
- Colour on different yarn bases — 46
- Stinging Nettles — 47
- To Dye with Stinging Nettles — 48
- Nettle, Pea and Watercress soup — 49
- Nettle Socks — 50

July — 53
- In the Dye Garden — 53
- Solar dyeing - Here's how you do it — 54
- 3 months have passed — 55
- Foraged dye this month — 55
- This is what I did — 56
- Crochet Hexagons — 57

August — 59
- In the Dye Garden — 59
- Tansy (*Tanacetum Vulgare*) — 60
- Amaranth as a dye — 60
- Jar 1. Water and Daylight method — 61
- Jar 2. pH3 and darkness method — 61
- A bit more about Amaranth — 61
- Orange Amaranth Bread — 63
- Amaranth and Garlic cheese crackers — 64

September **65**
- In the Dye Garden   66
- About Woad   66
- Dyeing with Woad slush   67
- Safflower *(Carthamus tinctorius)*   67
- Dyeing with Safflower   68
- Hopi Sunflowers   68
- Chamomile Cake   69

October **71**
- In the dye garden   71
- Dye supplies and Rhubarb Root   72
- Acid pH   72
- Alkali pH   73
- Iron Solution   73
- Copper Solution   73
- Cochineal   73
- Using Cochineal to dye with   74
- Loving the Lattice   75
- Red Cabbage with apples   78

November **79**
- In the dye garden   80
- More about Madder   80
- How to store natural dyes   81
- Newgale Beach   83

December **85**
- Gifts   86
- Natural dyes in liquid form   86
- Confetti dye effect   86
- What about dye courses?   87
- The Last month   87

Resources **88**
- Natural Dye suppliers   88
- Dye seed suppliers   88
- Books on my shelf   88

THE NATURAL YARN DYER'S ALMANAC

INTRODUCTION

This book has been written to encourage you to have a go at natural dyeing. It is not a complex, scientifically detailed book with reference to colour chemistry – there are plenty of those written by expert dyers.

I have written this book to teach you the basic methods and techniques of natural dyeing from my experience.

In 2017, my friend Steph, who is a yarn dyer, asked me if I would like to join her on a natural-dye course run by a local dyer, Amanda Hannaford. I decided to go along, just for fun, declaring that I didn't want to dye my own yarn – famous last words. I was hooked! The natural colours, techniques and simple fun inspired me and my journey began.

I am self-taught with a huge appetite for learning, reading and experimentation. After all, the worst that can happen is that you end up with a sludgy brown skein of yarn, or very little colour at all.

So, I learned, practiced and experimented, and over time my skills improved. Yours will too. I grew in confidence as I understood the ways dyes behave and how to get different colours and effects from the dyes.

The sustainability of natural dyes has always been a positive aspect of this fascinating artisan craft for me. Modern mordants mean there is no need to use toxic metals like tin and chrome to help fix the colours and so natural dyeing can be environmentally friendly and safe.

The dyes themselves and the dye plant products are safe to tip away on the garden, down the sink or put in the bin and they will do no harm. It's nice to think of dye plants growing in the compost that came from the used dye plants, that grew in the compost from dye plants.... You get the idea.

In this book, month by month, almanac style, I hope you will feel inspired to try some natural dye 'recipes'. We will explore colour in your store cupboard, forage for dye plants, grow dye plants, and look at dye stuff you can buy dried and ready to use.

I will explain about substantive and fugitive dyes and what this means when you choose what dyes to use.

I will explain what mordants are and when and how to use them; how to mordant protein fibres like wool and silk, and how to mordant cellulose fibres like cotton and bamboo. I will test the idea that soya milk can be used as a mordant on cellulose fibres. We will look at modifiers and pH to understand how they can subtly shift the original dye colour to another tone.

Perhaps you will have a go at growing your own dye plants, a bucket on a balcony or doorstep is enough space for a dye plant.

Lastly this book has a few recipes to enjoy, and it has knit and crochet patterns for things to make with your hand dyed yarns.

Sincere thanks to Amanda Hannaford for her technical editing of this book.

I hope you enjoy the natural dyes through the year with me,
Best wishes,

Caroline

GUIDELINES OF RESPONSIBLE NATURAL DYE FORAGING AND GROWING

- Know your environment and be able to identify any toxic, invasive or at-risk species.

- Know what you are foraging and collecting. If you can't identify it, don't pick it.

- If its toxic, pick carefully and cautiously, keep away from children and pets.

- Never take more than 20% of any plant, and only pick up windfall lichen, NEVER pick it from branches and trees.

- Be sure that you are allowed on the land you are foraging from. Keep to footpaths and don't cross fields or private land.

- Be careful where you plant – some dye plants like Madder and Woad can be very invasive and spread quickly.

- Share dye plants within your area, but only post plants or seeds if you are authorised to do so. Generally, don't share dye plants out of your own country.

HOW TO COLD MORDANT YOUR YARN OR FIBRE –

Mix the required amount of mordant (I use alum with cream of tartar) with warm water in a plastic jug. Pour this into a bucket and add more cold water, stir well to dissolve the alum and cream of tartar powder. If you are using a rhubarb-leaf mordant, pour the required amount of liquid for the weight of fibre into the bucket and add cold water. Tap water is fine to mordant with, but you can use rainwater if you prefer.

Un-twist the skein of yarn to be used until it's a large loop then tie it loosely with uncoloured string or yarn in several places. This helps prevent the yarn getting tangled and unmanageable. If you are using fibre, lay it on a surface and wet it until it is damp.

Put the yarn or fibre into the bucket of mordant solution and gently push it down to submerge it. Wool tends to float so it will need a bit of encouragement to sink. It's important to make sure that all the yarn/fibre is submerged so that it can absorb the mordant and become thoroughly wet. Gently stir the fibres taking care not to tangle them too much.

Leave the yarn/fibre to soak for 24 hours ideally. Lift the yarn/fibre out of the bucket and gently squeeze the excess water out. Rinse the yarn well to remove excess mordant. You can use the yarn/fibre straight away, still damp, or it can be dried and then used later.

HOW TO PREPARE A DYE

A lot of natural dyes are made in the same way as making a strong pot of tea. Usually, you soak the dry dye stuff in water in a pan then gently simmer for about 45 minutes to release the dye colour. The dye is then strained and used to dye yarn or fibre. The spent dye stuff can sometimes be used again to get more (weaker) dye colour, or it can be disposed of in the compost or household waste.

Some dyes can be used cold to dye the yarn, and some need to be heated again for maximum colour potential.

HOW TO RINSE YARN AFTER IT'S BEEN DYED

Wear an apron or old clothes and rubber gloves. Lift the yarn from the cold dye pot and gently squeeze as much dye as you can from the fibres. Make sure the drips go back into the pot, not onto you or the floor or work surface. Put the yarn into an old washing-up bowl. Put some cool tap water into the sink and lift the yarn from the bowl into the sink. Alternatively, put cool water into a washing up bowl and then add the damp dyed yarn.

Always use cool water when rinsing your yarn or fibre, be very gentle as the fibres can stretch or felt together. Avoid sudden temperature changes as this can felt the fibres too.

Gently move the yarn and pick out any remaining bits of dye stuff. There will be colour in the water; it will be any excess dye that hasn't fixed to the fibres. (If you have used a fugitive dye its likely that most of the colour will run out at this point.) Tip away the coloured water and add more, rinse again. Keep repeating this until the water remains clear. You can use a mild, pH neutral laundry liquid to freshen the yarn, but its not essential. Be aware that some wash liquids may alter the colour of your yarn slightly if they are not pH neutral and the dye is able to be colour adjusted with acid and alkaline solutions. You need to treat the yarn or fibre just like you would treat the finished item you intend to make.

Lift the yarn/fibre and gently squeeze it to remove excess water. Dry skeins by hanging up on a line outside if you can, or on an old towel on a radiator, not too hot. Dry fibres by blotting on an old towel and drying flat.

HOW TO EXHAUST DYE

After you have removed the yarn or fibre from the dye pot, there may still be colour left in the dye. This can be used to exhaust dye. Put more mordanted yarn or fibre in the cool dye and repeat the dye process as before. The colour will be paler than the first-time round, but will be a tone of the original colour. Some dye stuff exhausts well, and little colour is left, (Saxon blue) but some dyes are very 'generous' and you can get 3 or 4 exhaust colours. (madder and cochineal).

The dye stuff that is strained from the dye (either the pot or from a solar dye jar) can be reused. Prepare the dye pot in the usual way and use this dye colour too. The colour will be a bit paler but still has dye potential.

IMPORTANT HEALTH AND SAFETY INFORMATION

Although these dyes are 'natural', they still contain chemicals which may harm you.
Therefore, it is important that you follow some simple precautions so you stay safe and can enjoy this fascinating hobby without worry.

- Keep all equipment and utensils used for dyeing separate from food utensils, e.g. once a pan or spoon is used for dyeing, don't use it for food again.

- If you can, prepare and use your dyes in a separate space, away from food preparation, ideally outside or in a shed or garage with plenty of ventilation.

- Avoid breathing any natural-dye dust or fumes.

- Don't eat, drink or smoke when using natural dyes.

- Keep small children and pets away from your dye area.

- Dispose of used dyes and mordant solutions carefully. They can be put on the garden around non-edible plants, into the compost or in household waste or down the drain in small quantities. Please check with your local water company if you have a septic tank.

- Wear apron or old clothes and rubber gloves when handling dyes, mordants and any modifiers.

- Store dyes, mordants and modifiers out of reach of children in clearly labelled containers.

- Be sure you are aware of first-aid and emergency treatment if dyes, mordants or modifiers are accidentally ingested.

TELL ME MORE ABOUT THE HISTORY OF NATURAL DYES

Natural dyes have been used for thousands of years to colour textiles, paints and interiors. Possibly the earliest discovered use of natural red and orange dye colours was found dating back to 2,600 BCE. Recently a credible scientific study reported on the recent find of indigo dyed fabric dating back to about 4,200 BCE. Clothing was no longer limited to dull brown and grey. Colour could represent status and also became a statement of wealth with more rare dye colours costing more.

However, it was in the 17th century that the artisan skill of the professional dyer became recognised with legislation to ensure the quality, consistency and depth of colour required. Many 'recipes' for dyes had been developed by this time and were often closely guarded professional secrets. If you could dye woollen fabrics, silks etc to a consistently high standard you could command a high price.

The legislation and professional guilds at this time, especially in France and Italy, controlled the quality of the dyes, the cloth merchants and the dyers. Dyes which gave a 'good' colour were considered a higher quality than those which didn't. So how to qualify the term 'good' in 17th century terms?

WHAT CONSTITUTES A 'GOOD COLOUR?

A good colour had to have certain qualities. Historically, there were committees who would meet to preside over the dye colours and their merits.

Beautiful colours were expected – a deep colour saturation, and 'bright and lively' were common descriptions of a good colour. Of course, the more difficult colours to produce were nearly always in demand as good colours. As you might expect, the beauty of colour is subjective, and what was popular in one area may be different in another region.

A good colour might be fashionable at one time: deep blue, bright clear reds and turquoise, as well as pinks, purples and lime green that rose and fell in popularity throughout the century. In mid-17th-century France, colours with names like' liver-colour' and 'Paris mud' had a recognised place in the dye palette.

The workability of a dye also helped determine if it was good or not. It was not good if a skilled artisan could not work the dye to consistently produce the colour.

The permanence of colour was, and still is, of primary importance when using natural dyes. Good natural dyes need to be solid, lightfast and lasting, not altered by light or washing, not likely to rub or peel off over time. Importantly, these dyes do not harm the dyer or the consumer, as could happen with the use of lead- and arsenic-based colours.

Cost was a lesser but not insignificant consideration in the selection of good dyes. When dyeing textiles using a 1:1 ratio by weight of dye to fabric, this would have to be calculated. Local, cheaper dyes were used, but only if they gave a very similar colour with enough 'good' qualities to make the finished textile commercial.

Nowadays we use the terms 'substantive', 'adjective' and 'fugitive' when considering how good a colour might be for a dye project.

Substantive dyes are those which don't need a mordant to 'fix' the colour to the fibre. They are usually from plant sources that contain tannic acid (walnut husks) oxalic acid, (rhubarb roots).

Adjective dyes are those which require a mordant (such as alum) to help the colour fix permanently to the fibre. ('Mordant' comes from the French, meaning to bite or grip.) Fibres are pre-soaked in the mordant solution which opens the fibre and allows the colour from the dye to penetrate at a molecular level. These dyes are usually used at a 1:1 ratio of dye to fibre to give a strong clear colour. The addition of cream of tartar to the mordant solution helps brighten the finished colour. Adjective dyes are madder, calendula, safflower and lac insects.

Fugitive dyes are those which literally run away and hide. Even if used with a pre-mordanted yarn, beetroot, for instance, will not give a good, lasting or colour-fast colour as a dye. It tints rather than dyes the fibre and therefore washes away. The dye bath may look a delicious shade of cerise pink, but the yarn will rinse to a disappointing dull grey. Many of the berry and vegetable dyes are fugitive, despite what the internet tells you!

TERMS, TECHNIQUES AND A BIT MORE INFORMATION

Throughout this book there will be terms and phrases used that you may not be familiar with, so I thought it would be helpful to tell you a bit more before we start the year together:

NATURAL DYE

A natural dye is a colour which comes from the natural world and is not created chemically or synthetically. Natural dyes can come from plants, insects and minerals. These dyes can be used to dye protein or cellulose fibres such as fibre, yarn or fabric. A natural dye can be a powder, liquid, crushed insect, bark, flowers, roots, leaves or type of nut or fruit like acorns and walnuts.

Natural dyes are not always sourced from plants and not all plants give dye colour.

METAL SOLUTIONS

Metal solutions are created to adjust the colour of dyed yarns. They are usually copper or iron solutions and are made by dissolving copper sulphate crystals or ferrous sulphate crystals in water. Copper solution turns colours greener or browner in tone and iron (ferrous) solution tones down colour, they are said to be 'saddened'. Iron solution modifier also helps improve the colour fastness of dyes.

MORDANTS

Most natural dyes need a mordant to help them fix or 'stick' to the fibre and to slow down any fading. Mordants used to be made with toxic metals like chrome and lead, but now most natural dyers use safe mordants like alum (potassium aluminium sulphate). I use cream of tartar (Potassium Bitartrate) with the alum to brighten the colours. Both alum and cream of tartar are slightly acidic. Cream of tartar assists the absorption of the alum by the fibres and helps to ensure that little or no aluminium is left in the mordant solution when it is finished with.

There are natural mordants like tannic acid, which is found in oak galls and buckthorn bark, and oxalic acid, which is found in rhubarb leaves. Both these, in solution, will act as mordants to fix the colour of the dye. So a mordant can be made with oak galls or rhubarb leaves and used in the same way as an alum mordant. Of course, if the plant source you are using as a dye has tannins or oxalic acid, you don't need to additionally mordant with anything.

PH, ACID OR ALKALI

Some dyes are sensitive to pH, that means the acidity or alkalinity of the dye pot is important for the finished colour.

Acids, like vinegar (acetic acid) and lemon juice (citric acid) added to the dye bath will lower the pH to less than 7 and tend to shift colours to a more yellow tone. Reds become orange and purples become more red.

Alkali, like washing soda or household ammonia, will shift colours pinker or red, for example mustard-yellow rhubarb root can shift to coral red with the addition of small amounts of ammonia to the dye pot.

It is worth using a pH neutral laundry liquid to rinse your yarn/fibre to keep the colour you want. Some laundry liquids can be quite alkaline.

WATER

Depending on where in the country you live and on the source of your water, it will be classified as 'hard' or 'soft'. This means how much your water is affected by the minerals in the earth through which the rain water percolates before it gets to you. Hard water areas have lots of dissolved calcium minerals (more alkali) and this is often seen as a white chalky film inside your kettle. Soft water areas are often on peaty soil (acidic) and have few dissolved calcium minerals. The pH variation is generally very subtle but it might alter the base dye colour before it is modified in any way. This means that even if you follow a dye 'recipe' exactly, your results may be a subtly different colour tone to those of a friend who is a natural dyer who lives in another part of the country. Here in Cornwall the water is soft as it flows through the peaty soil of Bodmin Moor.

Protein fibres readily absorb natural dyes due to their structure. This helps protein fibres bond well with the colour molecules in dyes. The structure of a protein fibre is generally made up of a core surrounded by scales which 'open' up with the mordant and this helps the dye colour penetrate the fibre.

FIBRES

The structure of protein fibres (from an animal source) and cellulose fibres (from a plant source) are very different.

Protein fibres readily absorb natural dyes due to their structure of amino molecules (alkali), carboxyl groups (acid) and organic molecule groups. Opposites attract – acid to alkali, and the organic molecule groups are pH neutral. This chemistry helps protein fibres bond well with the colour molecules in dyes. The structure of a protein fibre is generally made up of a core surrounded by scales which 'open' up with the mordant and this helps the dye colour penetrate the fibre.

Cellulose fibres are characterised by a hard, waxy outer covering – think of how bamboo is. This outer covering has to be broken down mechanically, chemically or by heat to make fibres which can be used in textiles (retting). Often these cellulose textiles or yarns still have an element of this waxy coating which is resistant to dyes. For this reason, cellulose yarns and fibres need to be thoroughly scoured with heat and a chemical like washing soda as well as needing mordants to fix the colour.

Colours from natural dyes on cellulose fibres are often disappointing when compared to the same dye on protein fibres.

DYE VATS

Indigo, (Indigofera species) and woad, (Isatis Tinctoria) both give blue colours when used in a dye vat. This process is a totally different way of dyeing and doesn't fit with the water extraction methods used with most other dyes and here in this book.

GLOSSARY OF TERMS

CLASSIFICATION OF COLOURFASTNESS

Natural dyes can be classified as:
- Substantive (or direct dyes); these will fix colour to the fibre without the addition of a mordant. This is usually because they contain natural mordants such as tannic acid (buckthorn bark, oak galls, Persian berries). These dyes can be used with a mordant to enhance their light and wash-fast properties or to achieve different colours.

- Adjective (or mordant dyes); these dyes need a mordant to create a bond between the dyestuff and the fibres. Even with a mordant, dyes should be light and wash-fast tested and some dye stuffs perform better than others.

- Fugitive; these natural dye colours, even with a mordant, don't last. As the name suggests, they run away and hide! They wash away and fade quickly. They are the least stable and reliable dye colours, often from fruit, berries and flowers.

Most natural dyers just use the terms substantive, for the long-lasting colours, and fugitive for the fast-fading colours.

CELLULOSE/PLANT FIBRE

A cellulose fibre or yarn is one that is produced from a plant like cotton, linen, flax or nettle. It also includes modern fibres like Tencel© and Lyocell© which are produced from wood pulp, often eucalyptus.

DYE POT/PAN

Use an old stainless-steel or unchipped enamel saucepan for your natural dyeing, one that won't be used for food afterwards. Avoid aluminium and cast-iron pans because the metal will affect the colour of your dye. If you plan to dye more than 100g of yarn, you need to have a bigger pan. I use industrial-size stockpot pans and I can just about dye 1kg in those. So the size you need will be proportional to how much fibre you want to dye. I bought mine from an online supermarket, but boot sales, second-hand and charity shops are good sources for cheap pans and equipment.

EQUIPMENT

Most dye equipment is not technical and can be old kitchen pans, spoons etc. Remember, after using them for natural dyeing, that they should not be used for food again. You will need a large pan, plastic jug, several spoons, including a long-handled one to stir the dye pot, a set of small scales (i.e. ones that can weigh small amounts clearly), a sieve or colander, tongs, bucket and Kilner type jars.

EXHAUST DYEING

After the first amount of yarn or fibre has been dyeing in the dye pot, there is often enough colour to dye a second or even third amount of yarn. This is called exhaust dyeing. The colour will decrease as less colour molecules remain in the dye and give the series of yarns a pleasing ombre colour-fade effect.

FIBRE/YARN

These terms are used interchangeably as some of you will want to use this book to dye fibre for spinning or felting, and some will want to dye yarn like I do.

FUGITIVE

Dyes are said to be fugitive when they run away and hide. What this means is that the dye colour won't 'stick' to the fibre, even if you use a mordant. As you rinse the yarn after dyeing, the colour will rinse off and run away, any remaining colour is likely to fade quickly too. Berry and vegetable dyes are often fugitive.

MODIFIER

A dye modifier is a chemical solution that will alter the tone of the dye colour or change it completely. Modifiers may be acidic or alkaline, or made from solutions of metals like iron or copper.

MORDANTING

I prefer to cold mordant by mixing specific amounts of alum and cream of tartar in a ratio to weight of fibre (WOF) with hot water in a jug until dissolved, then adding to a bucket of lukewarm water and submerging the skeins of yarn for at least 12 hours, and up to 24 hours. Some natural dyers prefer to mordant in a pan with heat. This is done by mixing the alum and cream of tartar with water and stirring well. Wetted yarn is then added and the pot is simmered for an hour. The yarn is left in the pot overnight or at least 4 hours, moving the skeins from time to time. The yarn is lifted from the pot and rinsed. I have little experience of using the hot-mordant method as I find the cold-mordant method satisfactory, and much quicker.

PREPARING THE DYE

You need to infuse the dye stuff in a large pan of warm water for at least 30 minutes to release the colour. Most dyestuffs benefit from simmering for between 30-60 minutes before straining. The general rule is that the tougher the plant material is, the longer the simmering needs to be. Really woody, tough dye stuffs benefit from soaking in cold water for a couple of days to soften first.

Once the dye has cooled is it strained ready for use.

Alternatively, the yarn can then be added to the pan with the dye stuff still in place (some dyers find this gives a stronger colour). Some dye stuffs are better heated to fix to the yarn, some can be used warm or cold. Used dye stuff can be saved and often reused for a second dye colour. The dye liquid can also be reused to dye a second batch of yarn a paler colour.

Some dyes need to be prepared by soaking in an ammonia solution, for example some lichens, and some dyes need to be prepared and soaked in the dark (e.g. amaranth).

Lichens can give colour by the simmer method described above and give colours ranging from beige, through golds to deep, tan brown. Lichens used for pinks, reds and purples have to be fermented in ammonia. The only one that is easy to get hold of, (particularly in Cornwall) is Evernia Prunastri. This is the most common windfall lichen.

For anyone wishing to know more about lichens and dyeing with them, I recommend contacting a lichen specialist to help confirm identification before proceeding. Lichens are a very ecologically important and sensitive group of organisms and should not be picked in the countryside, only picked up from windfall branches etc.

PROTEIN FIBRE

A protein fibre is one that comes from an animal source. For example, wool, alpaca, cashmere and silk.

SCOURED FIBRE/YARN

Sheep wool, in particular, has a lot of lanolin which helps make it waterproof for the sheep. This lanolin will make the wool resistant to water-based dyes too, so wool and other animal fibres need to be cleaned and prepared. This is called scouring. It involves cleaning the fibres with alkaline soap to remove the lanolin then rinsing and drying. This is done before spinning the fibre into yarn.

SOLAR DYEING

This is a technique of dyeing protein yarn or fibre in a large, lidded jar over a length of time, exposed to UV in the daylight. Varied and colourful effects can be created with this simple method which would be difficult to achieve in a traditional dye pot.

TINCTORIA

This Latin term in a plant's name gives a clue that the plant may give a dye colour, for example Anthemis Tinctoria is dyers chamomile, Rubia Tinctoria is madder root and Alkanna Tinctoria, dyers alkanet.

TRADITIONAL DYE METHOD

The traditional method of preparing dye was in large vats or 'baths'. We can use a pan/pot for natural dyeing at home (as described above – 'Preparing the dye').

SUBSTANTIVE

Dyes are called substantive when they fix without the need for a mordant because they contain a mordant naturally. However, their light and wash-fastness can be improved by using an additional mordant.

WOF

This abbreviation stands for Weight Of Fibre and is used to indicate the ratio of fibre to another substance in the dye process. For example, mordant: WOF is 10%. This means that if you have 100g of fibre/yarn you need 10g of mordant.

PLANTING AND GROWING GUIDE

MADDER

Madder gives a range of colours from brick red through to orange, peach and pink. It is best grown from root cuttings as seeds are reluctant to germinate. Plant root cuttings after the risk of frost in late spring. Madder does well in most soil, but prefers alkaline/chalky soil for the strongest red colours.

To harvest: dig up the thickest root cuttings 3 – 4 years after first planting (for the richest colours) These can be used fresh, or chopped and dried for use later. According to Jenny Dean, (see books on my shelf, at the end of this book) using madder tops, after they have died down for the winter, can also give good colour.

COREOPSIS

This bright flower gives orange dye colour. Sew the seeds where they will bloom after the last frost. Coreopsis prefers fertile, well-drained soil.

To harvest: Pick flowers in full bloom on a sunny day and dry for later use.

SAFFLOWER

Safflower can give both yellow and red colours. Plant the seeds in trays in March/April then harden off before planting out in well-drained soil.

To harvest: Pick a few stamens at a time and dry. Protect the flower heads from the birds who will pick at the seeds.

SANGUINARIA/BLOODROOT

Like madder, it gives red and orange tones. Grow from root cuttings and harvest in the same way as madder. It prefers semi shade and clay/loam soil.

CALENDULA/POT MARIGOLD

Calendula/pot marigold gives shades of yellow. Sow seeds in situ in the autumn or spring, or sew in trays and plant out. It prefers a sunny spot and makes a pretty border flower.

To harvest: pick flowers in full bloom on a sunny day and dry for later use. Can also be used fresh.

DYER'S CHAMOMILE

This reliable dye plant gives rich yellow colours. Grow from seed in trays from March, or sew direct from May to July. Thrives in a sunny place in the garden.

To harvest: pick flowers in full bloom on a sunny day and dry for later use. Can also be used fresh.

DYER'S GREENWEED

Another dye plant that gives yellow. Plant seeds in trays in March/April and transplant to the dye garden in June. It prefers slightly acidic soil and slugs love the seedlings.

To Harvest: pick young shoots for the best colour. This plant can become quite shrubby.

GOLDEN ROD

Golden rod also gives yellow. Sow seeds in trays in early spring. The plants will form a perennial clump and it does well at the back of a sunny border.

To harvest: Cut whole flower spikes and dry for later use.

MULLEIN

More yellow dye…. This biennial plant should be sewn from seed in October where they are to flower. Use Soda ash when dyeing with mullein to get stronger yellows.

To harvest: Cut and dry flowers and leaves to dye with.

WELD

And another yellow dye plant. Plant seeds in trays in the autumn, they will germinate in the spring. Plant in a dry, sunny spot.

To harvest: Cut the whole flowering spikes and dry for later use.

TANSY

Tansy gives greenish yellows which over dye well with Saxon blue for lovely greens. Plant seeds in trays in the autumn or spring, and plant out in late spring when the seedlings are big enough to handle. Tansy can cause skin irritations when fully grown, so wear gloves when handling these plants.

To harvest: cut flower clusters as they open and dry. They will look like little yellow buttons, don't worry about separating them individually.

JAPANESE INDIGO

Use fresh seeds on the surface of the compost in a tray in March. Plant out in rich soil and water well. To harvest: cut leaves before the plant flowers. Use secateurs and discard old leaves.

WOAD

Plant seeds in trays in March and transplant to the dye garden when the risk of frost has passed. To Harvest: as for Japanese indigo, cut all the leaves as this allows for new growth

HOPI SUNFLOWERS

Plant seeds in individual pots from mid-April after the last frosts. Protect the flower heads from the birds. To harvest: wait until the flower head has bloomed fully and is beginning to fade. Cut the flowering head and remove the seeds with a fork.

JANUARY

Poem for January

Cold January
When early bright blue frosts dance
In spite of the sun

—

January regularly has frosts, ice and snow and is the coldest month of the year.

Average temperatures for January: low 5° high 9°.

Average 15 days of rain.

January was named after the Roman god Janus, who is depicted with two faces, looking forward and backwards. This symbolises beginnings and endings, in this case of one year and the next. Romans called it Januarius, even though their new year began in March.

Anglo Saxons called the first month of the year 'Wolf monath' because that's when the wolves would come into the villages looking for food.

Traditionally, the first of January is a significant day regarding prosperity and wellbeing in a household. On farms, a flat cake was put on a cow's horns, and if it fell off in front of the cow that meant good luck for the year. People would visit other households with gifts of coal for warmth and bread so everyone would be warm and have enough to eat in the year to come, and gifts of money and greenery for a long and prosperous life. Wassailing, or the singing of traditional folk songs, has been a way of passing on good wishes for the year ahead from as far back as 1400's.

JANUARY IN THE DYE GARDEN

January is a quiet time in my dye garden. Its normally too wet and cold to get out and dig over the ground or deal with the plants in tubs. It's much better to stay indoors with a hot cup of tea and a cinnamon bun or piece of turmeric cake and dream of what dye plants to grow this year.

It can be difficult to know how to start a dye garden and a lot depends on what you want to dye. I enjoy dyeing spun yarns, mainly animal fibres, but sometimes plant fibres too. The dye plants I grow are destined to colour these yarns and although you can use them to dye fabrics, I have little comprehensive experience of this.

My joy is with the beautiful colours dyed onto yarns which are then knitted or crocheted into wearable items.

So have a think about what you want to dye, and then what sort of colours you want. If you are looking for strong, bright colours then some, but not all, natural dyes give you these. In the main natural dyes are more subtle and gentle.

CHOOSING AND BUYING SEEDS

I like to choose and buy my dye plant seeds in January. If feels like a sunny thing to do when the days are short and dark. It makes me feel that spring is on the way.

There are very good, specific dye-seed suppliers and you can find them with a quick internet search. (See suppliers list at the back for a few UK ones.)

Be sure that you buy seeds that are specifically for their colour potential (which is why it is better to go to a dye-seed supplier). For example, *Anthemis nobilis*, chamomile, is not the same as dyer's chamomile, *Anthemis tinctoria*. The 'tinctoria' is a clue, and usually means that colour/tints/tincture can be extracted from the plant. That's not to say that flowers and plants can't give colours to dye with, but the 'tinctoria' and specific dye plants give the best intensity of colour.

Buy your seeds from a reputable supplier – one who gives advice on how to grow and tend the plants. Dye seeds can be tricky to grow and it's useful to be able to refer back to the supplier if you need advice.

WHAT NEXT?

This year will be an adventure I'm going to share with you: the highs and lows of growing dye plants from seed; the colour extraction methods; the successful dyes and less successful ones; the use of mordants and modifiers – and a cup of tea or two along the way.

Some dye plants like madder, *Rubia tinctorum*, (there's the tinct- bit again) and bloodroot *Sanguinaria canadensis* are easier to grow from root cuttings. You can buy these online or ask a dye friend. Madder can be an invasive plant once established so dye friends are usually happy to share! It is possible to grow madder from seed but it is a reluctant germinator.

Many dye plants are annuals so you will be able to use your home-grown dye colours this year. Others are biennials or perennials and you will have to wait (Refer to the planting and harvesting guide.)

THE COLOURS

A lot of dye plants give tones of yellow. They are subtly different and I don't like to limit myself to just one or two of each primary colour. The tonal variations you can get from a range of, say, yellow dye plants are stunning, subtle and naturally beautiful. From pale yellow from dandelions (Taraxacum) and gorse flowers (Ulex) to dark bright yellow of chamomile (Anthemis tinctorum) and mustard yellow from edible rhubarb roots. (Rheum rhubarbarum or rhaponticum). Several other ornamental plants of the Rheum species will also give similar results.

The intensity of colour can vary depending on what fibre you are dyeing onto. Chamomile dyes beautifully onto animal fibres like wool or alpaca but is disappointing on plant fibres like cotton or linen.

So, you have an idea of what dye seeds/colours you want to grow…. But where will you be growing your dye plants?

This year I am going to experiment and grow my dye plants in large tubs. This is mainly because we have a problem with the invasive weed comfrey and I don't like using weed killer. It's tricky to pull up its thick roots from among the dye plants and any bits of root which break off will grow again. You see my problem? So, I'm going to live with the comfrey and keep my dye plants in tubs. Of course, you can grow your dye plants in the ground if you have the space and patience to pull up the weeds. I don't!

The advantage of growing dye plants in pots is that the soil and growing conditions that suit each variety can be more easily managed and harvesting is simpler too. Madder has always been in a pot in my garden because it's a take-over thug! Growing plants in pots also means that even if you only have a patio area or balcony you might be able to grow a dye plant variety or two. You don't need a big garden or allotment to grow your own dye colours.

The disadvantage of growing anything in a pot is that it takes more tending, more feeding with an organic fertiliser and regular watering. However, this is no hardship; watering plants with collected rainwater on a sunny evening is a real pleasure to me.

I re-use seed trays from year to year. Just give them a thorough wash with soapy water before you fill them with compost and begin to sow. It's a job for January if you have a nice day and want to be outside in the fresh air. There's not much else to do this month but wait… for your seeds to arrive, to plan where your plants will grow and thrive best and to get ready for your new project.

If this is your first year of growing dye plants you may not have any dry dye stuff stored ready for use yet. Why not raid your kitchen cupboard for natural colours? They may not last long as a dye colour but are fun to use and try.

DYEING WITH TURMERIC

Many cultures have used turmeric (*Curcuma longa*) as a dye stuff over hundreds of years. It has been used to dye fibres and fabrics, baskets and woven items, even skin, and as elephant paint. Predictably, it gives a bright yellow colour to everything it touches. You have been warned!

Turmeric, and some other natural dyes are not lightfast which means they will fade when exposed to UV light. They are said to be 'fugitive'. The fade can be slowed by using a 'mordant' which helps to bind the colour to the yarn for a bit longer. Alternatively, you could re-dye your finished item if you want to refresh the colour. You might like the fade….

TO DYE YOUR YARN WITH TURMERIC:

You Need:
- 100g undyed yarn skein or clean/scoured wool tops, you can add scraps of undyed raw silk fabric too. Tie your skein of yarn with loops of string in several places to prevent tangling.
- 15g powdered turmeric
- 10g alum (See section on mordants for more information)
- 4g cream of tartar
- Non-aluminium saucepan
- Colander and bucket
- Gauze/muslin fabric
- Rubber gloves
- Apron
- Eco -pH neutral laundry liquid soap

Feel free to multiply up the yarn/fibre quantities, just multiply up the mordant and turmeric by the same amount/ratio.

1. Soak the fibres in fresh water overnight in the bucket. This enables the fibres to take up the mordant evenly which in turn helps achieve a more even dye colour.

2. Mix the alum and cream of tartar mordant in warm water in a jug until dissolved then pour this gently into the bucket with the yarn and stir.
3. To make the dye, simmer 15g of Turmeric/100g yarn in about 500ml of water for about an hour so the turmeric disperses. Leave to cool.
4. Empty the yarn and mordant from the bucket into the colander to drain then lift out and leave to one side, still damp.
5. Line the colander with the gauze then, wearing rubber gloves, strain the turmeric solution through the gauze into the bucket. (The gauze will be stained by this process. You may need to repeat the process with a coffee filter if the gauze is not fine enough but it's not a problem if you don't. You will just get a fine layer of undissolved turmeric on your yarn which takes more rinsing off.)
6. Place the mordanted yarn into the saucepan and pour on the turmeric dye from the bucket. You may need to add more water to cover the yarn. Gently simmer for an hour. The heat needs to be gentle and slow; sudden changes of temperature can felt animal fibres like wool. After an hour, leave the fibres to cool.
7. Gently lift the yarn out of the pan – don't throw the dye away yet – and gently rinse the fibres in lukewarm water until the water runs clear and colourless and the dye has fully rinsed away. Use a pH neutral eco-laundry liquid soap to help with rinsing. (Non pH neutral soaps can temporarily change the colour of the yarn to red.)
8. You may be able to use the dye water a second time to get a paler yellow colour. This is called exhaust dyeing. Just repeat the process as before with more mordanted yarn but omit stages 3 – 5 as you already have your dye.
9. Dry gently away from strong heat and out of direct sunlight. I try to hang yarn on the line to drip dry. If the weather is wet, I put yarn on old towels on a heated airer to dry.

THIS MONTH'S PATTERN

Now you have your very own skein of hand-dyed yarn. Here's a pattern using 4ply yarn that you might like to make....

Ariel Shawl

By Caroline Bawn / Gorgeous Yarns

Project description

This quick knit features simple stocking stich knitting, and an optional simple crochet lacy edge.

You will need...

100g 4ply yarn 3.5mm knitting needles and 4.0mm crochet hook.

Tension

In the range of 26st x 36 rows = 10cm. Finished shawl measures 130 cm wide x 30 cm deep, excluding crochet edge.

Method

Cast on 10 stitches and work 2 rows stocking stitch. (Knit 1 row, Purl 1 row)
Row 3. Cast on 3 st at the beginning of the next row and purl these 3 st then K the rest of the stitches in the row.
Row 4. Cast on 3 stitches at the beginning of the next row and Knit these 3 st then P the remaining of the stitches in the row.

Repeat rows 3 and 4 until you have 310 st.
Work 5 rows St st, with 5 P st at the beginning and end of the K rows.
Work 5 rows of knit (Garter st).

Cast off. If you want to crochet the lacy edge, don't cut your yarn.

To crochet the edge, work along the shaped edge only. D/cr into every stitch along the edge, turn,
D/cr into the first st, ch 2, miss 1 st,8 repeat * to ** to end of the row, turn,
Ch 3, *D/cr into the ch loop, ch 3**, repeat * to ** to end of row, turn,
Ch 4, *D/cr into the ch loop, ch 4**, repeat * to ** to end of row, turn,
Ch 5, *D/cr into the ch loop, ch 5**, repeat * to ** to end of row, ch 5 D/cr into last st of previous row, cast off.

Sew in yarn ends.

Abbreviations

Knit
K – knit
P – purl
st(s) – stitch(es)
m – metres
yds – yards
cm – centimetres
in – inches
mm – millimetres
g – grams
LH – left hand
RS – right side
WS – wrong side
Rep – repeat

Crochet
Ch – chain
D/Cr – double crochet

Cast on at start of the row by inserting the right-hand needle between the last 2 st on left needle, yarn round needle, pull loop through and put onto left needle. This is 1 st.

ABOUT TURMERIC

Turmeric (*Curcuma longa*) is the root of a plant in the ginger family. It is native to Asia and is grown primarily in India. It is traditionally used in Ayurvedic and other Indian and Chinese medicines, primarily as an anti-inflammatory, for skin problems, respiratory infections, joint pain and stomach disorders. It is most well known as an ingredient of curry powder but has gained popularity as an ingredient in its own right.

The colour in turmeric is Curcumin, a bright yellow, hence its name: 'Indian saffron'. The yellow spots on silk bandanas in the 19th century came from turmeric.

The pH in soap can change turmeric yellow to red but it returns to yellow again as it dries.

Turmeric is well known in middle-eastern cookery, including this delicious cake from Lebanon.

Sfouf Cake

Ingredients:
- 1 ½ cups semolina
- 1 ½ cups plain flour
- 1 tbs turmeric
- 1 ½ teasp baking powder
- ½ cup sunflower oil
- 1 cup milk
- 1 cup sugar
- 1 - 2 tbs tahini - sesame seed paste (can be replaced with oil)
- Handful of pine nuts or almonds optional

Method.
1. Preheat the oven to 190° C and grease a 9-inch baking tin with the tahini or oil.
2. Mix the dry ingredients together in a large bowl (semolina, flour, turmeric, baking powder).
3. Mix the wet ingredients (oil, milk, sugar).
4. Combine the wet and dry ingredients and mix to a smooth batter. Pour the batter into the prepared baking tin and sprinkle on the pine nuts/almonds if using.
5. Bake in the pre-heated over for 30-35 minutes until firm and until a skewer inserted comes out clean.
6. Cool on a wire rack and cut into 16 slices.
7. Store in an airtight container.

Delicious as it is, or with crème fraiche or thick yoghurt.

FEBRUARY

Poem for February

Chilly February
A golden daffodil waits
In the frozen air

The name February, the second month of the year, has only been in existence for about one hundred years. The name comes from Februa which means cleansing or purification, reflecting the rituals performed before spring. In Shakespeare's time the month was called Feverell and 300 years ago was called Februeer.

February 2nd is Candlemas day, the mid-point of the winter, halfway between the shortest day (21st December) and the spring Equinox (around March 20th). This was the date when all the candles for the coming year to be used in the Christian church were blessed, hence 'Candle Mass'.

Traditionally, Shrove Tuesday is in February and marks 40 days before Easter in the Christian church. Lent started the next day and so all the luxury foods in the house were used up beforehand. They are now replaced by pancakes.

Of course, Valentine's Day is 14th February and is associated with many romantic superstitions including:
- The first man a girl saw on 14th February would be her future husband.
- If a woman saw a sparrow, she would marry a poor man and be very happy, if she saw a robin, she would marry a sailor and if she saw a goldfinch, she would marry wealth.

The average high temperature in February is 9° and the average low temperature is 4° with 12 days of rain.

FEBRUARY IN THE DYE GARDEN

Did January seem like a long month to you?
I did manage to get into the garden a few times to do some tidying but it has rained nearly every day and everything is so wet.

My dye seeds have arrived and I've tucked them into the right spaces in my diary so I can keep on track with planting. Next month will be a busy month with a lot of varieties of seeds to plant, including the dye seeds.

POTS

I've been thinking about what large pots/tubs to plant my dye plants in this year.
Should I blow the budget and invest in terracotta but risk them cracking with the frost? (Despite what the labels might say about frost-proof pots, I have found that they crack, even in mild Cornwall.) Should I go low-budget and buy large plastic pots which will do the job but not look so lovely and of course not be in line with the eco-conscious lifestyle I'm trying to live?

I will keep looking. I'm hoping to find recycled plastic pots which will fit my budget, look OK and still be more eco-friendly than first-generation plastic.

Thinking about eco-credentials and sustainability is all part of why I use natural dyes. There is so much evidence about the harm that chemical dyes do if not disposed of thoughtfully. Even as a solo indie-dyer I don't want to contribute any nasties to our environment if I can help it.

MORDANTS

The word mordant comes from old French, 'Mordre', which means to bite or grip. That's exactly what the mordant helps the colour to do to the fibre.

Some natural dyes need mordants to 'fix' the colour to the fibre. I use Alum (Potassium Aluminium Sulphate) as a mordant and Cream of Tartar (Tartaric acid) to brighten the dye colour on animal fibres (wool, alpaca, silk etc). On plant fibres (linen, cotton, bamboo) I use Al-Ac (Aluminium Acetate) as a mordant and sometimes use chalk (Calcium Carbonate) as a binder or afterbath.

So, let's talk about the mordants I use.

Alum mordant and Cream of Tartar

ALUM

Alum is a naturally occurring metallic salt which is mined. However, following the Pope banning exports of alum to Britain following the reformation, the English discovered how to manufacture it in the 17th century. Alum nowadays is made in a factory rather than mined. It is a common substance found in products as diverse as pickles, deodorants and toothpaste. (Think of a 'rock crystal' deodorant.) As a mordant it has a negligible environmental impact when used and poured away. I always use food-grade alum because I figure if you can eat it, it won't do you any harm as a dye mordant in minimal quantities on your skin. Generally speaking, alum as a mordant is 10% of the dry weight of the fibre (WOF).

Similarly, cream of tartar is commonly used in baking with bicarbonate of soda to produce carbon dioxide bubbles to make your cake mixture light. Cream of tartar is a by-product of wine production, and is used at 7% WOF.

For plant fibres, the main mordant is Al-Ac, aluminium Acetate. This mordant is usually industrially produced. It is used at 5-10% WOF – the higher the % the darker the shade of colour obtained. Once the fibre has been mordanted in Al-Ac it is treated in a bath of wheat bran or calcium carbonate.

HOW TO MORDANT YOUR FIBRES

COLD MORDANTING

For animal fibres:
Soak your prepared fibres (clean, scoured etc) in a bucket of water overnight.

Mix the weighed alum and Cream of Tartar in 500ml warm water and pour into the bucket with the fibres. Stir gently to mix well. Leave for at least 30 mins. The fibres in the bucket need to be soaked with the mordant solution and can stay in the bucket for several days before use. Alternatively, the fibres can then be dried and saved for use later. The quality of the mordanting is not affected by drying the fibre. Longer soaking in the mordant does not seem to greatly affect the dye take up of the fibre, nor the intensity of the colour.

HOT MORDANTING

For plant fibres:
Soak the scoured fibres in a bucket of water overnight.

Mix the weighed Al-Ac (aluminium acetate) with 500ml hot water, being very careful not to breathe in the acidic fumes. Pour this solution into a large dye pan with the fibres and stir well to mix. Leave for several hours.

Either: wrap the wheat bran (5% WOF) in a piece of muslin (so your yarn doesn't get bits on it) and place in a large jug or bowl of hot tap water. The water will go milky, and you need to squeeze the bran parcel to extract more 'goodness'. Pour the milky liquid into the pan with the fibres and leave the yarn to soak in this milky water for at least 60 mins, then its ready to use.

Or: use calcium carbonate powder (5% WOF) in the same way as the wheat bran. Rinse the fibres well after this solution to correct pH.

Then simmer the yarn in the pan for about 60 minutes. Allow to cool before removing the yarn and rinsing well.

The use of mordants to fix the dyes is a matter of personal choice. If you want your yarn to be as truly natural as possible you can dye without mordants but the dye colours will be less strong and will fade more quickly.

There has been a lot written about natural mordants like soya milk. They are not mordants in the true sense in that they do not fix the colour to the fibre like alum or Al-Ac do. The soya milk acts more like a binder and does not give the intensity or stability of colour like alum or Al-Ac. Soya milk is more commonly used with artisan fabric dyeing to increase absorption of colours onto the fabric and prevent wicking where colour is painted onto the surface of the material.

Here are the results of some mordant experiments I did on wool and cotton yarns, all dyed in the same dye bath with madder:

From L to R. Cotton with Aluminium acetate, Cotton with soy milk, Wool with Alum, Wool with soya milk,

The colour of the madder on the wool with alum and the cotton with Al-Ac is 'true' and what I would expect. The colour of the madder on the wool with soya milk was a surprise as I didn't expect it to work well and thought it would be more like the cotton yarn with hardly any colour at all. So, the soya-milk mordant worked fairly well on the wool, although not as well as alum. Perhaps it was to do with the protein in the soy milk; I don't know.

Rhubarb leaves can be used as a natural mordant because they contain oxalic acid. More about this in March.

Some dye plants naturally contain their own mordants in the form of Tannic acid or Oxalic acid. They are said to be substantive. These are often tree-based dye stuffs like buckthorn bark, oak galls, walnut husks and alder cones, also rhubarb roots and leaves. If a minimum or zero environmental impact is important to you, these are the natural dyes to go for as they don't need a mordant like alum or Al-Ac.

Pre-mordanted fibre can be dried and stored for later use. It can also be over-dyed without the need for more mordant even if both dyes need a mordant. For example, this skein has been dyed yellow with rhubarb root and then overdyed with madder.

As there are no dye plants ready at this time of year, I'm still using dye stuff from the cupboards and larder. They are not light-fast so they will fade, but are fun to use and enjoy while the colour lasts.

Rhubarb root and madder over dyed yarn

AVOCADO DYE EXPERIMENT

I was taught that avocados are fugitive and don't give good colour. I am intrigued by the lovely rich pink shades that other natural dyers on the internet seem to get from avocados. So, this is an experiment which challenges what you see on the internet.
I have been saving avocado shells and stones for some time in a bag in the freezer. I've decided to use them now and I'm trying again. I must confess this is my fourth attempt. I have never managed to get anything like a rich pink, only a pale skin tone/peach.
   So, if you like avocados, a lot, then have a go. Here's what I did....

Dyeing with Avocado shells and stones

YOU NEED

- 40 avocado shells (1/2 avocado = 1 shell)
- 20 avocado stones
- 100g unmordanted wool yarn

METHOD

1. Simmer shells and stones in 2l of water for 45 minutes, cool, then repeat.
2. Drain the liquid through a sieve lined with muslin/cheesecloth into another dye pot, add just the stones and simmer for another 45 minutes.
3. Allow to cool and remove the stones.
4. Soak your yarn in a bucket of water with a drop of eco pH neutral liquid soap for at least 1 hour until the fibre is soaked. Lift from the water, squeeze out excess water and lower into the cool dye pot.
5. Gently heat the dye pot for 45 mins at 60° – no hotter or the wool will felt. Leave to cool completely before rinsing the yarn in cool water. Rinse until the water runs clear and no more dye colour comes from the yarn.
6. Dry naturally in fresh air, or with gentle warmth on a towel on top of a radiator.

The result is a barely there nude/pale skin colour. Very pretty but definitely not a strong pink.

DYEING WITH BLACK BEANS

A store-cupboard dye that works well is black beans/turtle beans.

The dry bean is soaked and the colour from the skin of the beans is extracted. Some dyers have got blues and violet colours. I have only ever got shades of grey. As a neutral contrast colour of yarn these can work very well so I do keep using black beans to dye with.

HERE'S WHAT YOU DO

Soak 1kg of beans in large pan with 10l of water. Stir every now and then. At least 2 hours before you plan to use the dye, stop stirring. The water will look a lovely purple shade.

Very gently skim the coloured dye water from the surface of the pot, being careful not to disturb the beans or sediment in the bottom of the pan. Pour this dye water into a new pot.

Use damp alum mordanted animal fibre (black beans don't dye plant fibres very easily) and add them to the dye pot. You can leave the yarn for a couple of days in the pot to absorb the colour, or heat gently at 60° for 30 minutes. Allow to cool before rinsing.

Be aware that the pot may start to get smelly and ferment a bit. This won't affect the colour of the dye or the yarn, but I wouldn't eat the beans if it does!

The remaining black beans can now be used again. This time add water to the soaked beans, simmer for about 2 hours and a grey cloudy dye will be produced. Allow this to cool and use as above with mordanted yarn.

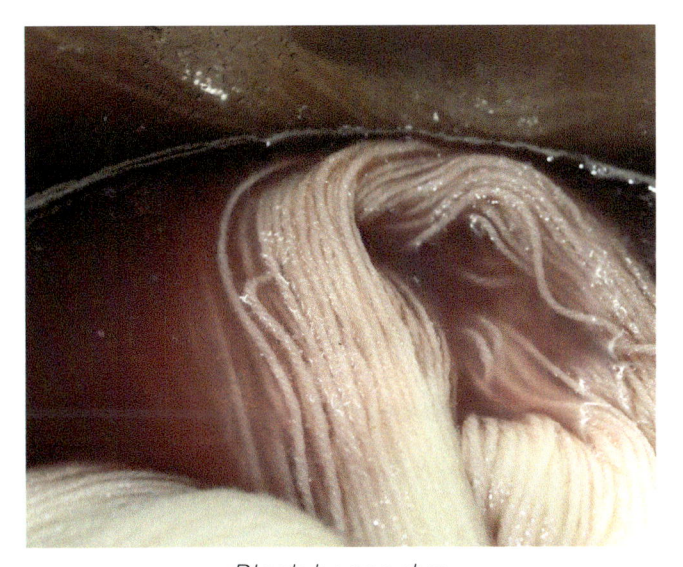

Black beans as a dye

Black beans dye.

THIS MONTH'S RECIPE

The black beans can be composted, but can also be eaten in a chilli. Here's a super quick and easy recipe....

Black bean chilli

Pre-heat the oven to 180°C

Ingredients

- 2 tbs Olive Oil
- 4 cloves garlic
- 2 large onions
- 3 tbs paprika
- 3 tbs ground cumin
- 2 tbs sugar
- 400g chopped tomatoes (canned is fine)
- 400g soaked black beans

Method:

1. Heat the olive oil in a large pan, and fry the onion and garlic until soft. Add the spices for a few minutes, then add the remaining ingredients.
2. Pour into an oven-proof dish, with a lid and cook for 45 minutes.
3. Serve with rice, crumbled feta cheese, chopped spring onion and chopped avocado.
4. Add sour cream if liked.

Pattern: a simple stash-busting Cowl

Here's a pattern idea for you to make with your Black Bean dyed yarn and left-over bits of yarn from other projects:

A simple stash busting cowl for men and women

You will need...

100g of 4ply and small amounts of contrast colours and 3.5mm circular 50cm knitting needles OR
100g DK yarn and small amounts of contrast colours and 4.00mm circular 50cm knitting needles, a yarn sewing needle

Tension

4ply worked over stocking stitch with 3.5mm needles: 38 st x 34 rows = 4inch/10cm
DK worked over stocking stitch with 4.0mm needles: 24 st x 31 rows = 4inch/10cm

Method

4ply instructions in BLACK, DK instructions in RED
4ply: Cast on 160 stitches in MC using a fairly loose or stretchy cast on, join into a round, PM
DK: Cast on 160 st in MC using a fairly loose or stretchy cast on, join into a round, PM

Work 12 rows in the round of stocking stitch. (In the round this is knit every stitch.)
Then *work 4 rows of k2 MC, K2 CC, carrying yarn behind loosely
Then work 4 rows MC**
The pattern is from * to **, Repeat 8 times for 4ply, and 6 times for DK.
Then work 12 rows in MC.

Cast off using a stretchy cast off. I used a Russian cast off.
The edges will roll over onto themselves, that's OK!

Russian Cast-off method is on our You Tube Gorgeous Yarns channel.

Abbreviations

K – knit
st(s) – stitch(es)
Rep – repeat
MC – main colour
CC – contrast colour
PM – place marker

MARCH

Poem for March

Winter slips away
Sheep in the fields like clouds,
March winds sing out, waking Spring!

March was originally the first month of the Roman year, named after Mars, the god of war. The calendar was only changed to the Gregorian style we know today in 1752, March becoming the third month, and January 1st being the start of the year.

The Anglo Saxons called March 'Hlyd Monath' which means 'Stormy month'. March traditionally has strong winds, due to the transitioning weather to the south-west as the gulf stream moves.

Average temperatures range from 10° high to 3° low with an average of 16 days of rain.

IN THE DYE GARDEN

At last, it is warm enough to start planting my dye seeds! Last weekend I set out my seed trays and spent a happy hour planting. Already, 4 days later the dyer's chamomile is showing seedlings.

I HAVE PLANTED

Dyer's chamomile *(Anthemis tinctoria)* for Yellow
Dyer's greenweed *(Genista tinctorum)* for yellow

Dyer's Chamomile seedlings

Woad *(Isatis tinctorum)* for blue
Japanese indigo *(Persicaria tinctorum*) for blue
Golden rod *(Solidago)* for greeny yellow
Safflower *(Carthamus tinctorium)* for orange and pink
If they all grow, I will have a lovely range of home-grown dye colours.

MORE MORDANTS

Continuing with the mordant experiments – this month did a trial with rhubarb leaf mordant. I must say I was surprised by the results because I expected the oxalic acid to change the way the madder showed on the yarn in the same way as changing the pH does.
I used the following method:
1. Harvest the leaves, wearing gloves, because the leaves contain oxalic acid, which as well as being a mordant is toxic in large quantities.
2. Weigh the leaves and note the weight in grams. Mine was 1425g
3. Chop up the leaves and put in a large non-aluminium, non-food, saucepan. (Do not use this saucepan for food again.)
4. Cover the leaves with hot tap water. Cook on the hob in a well ventilated or outdoor space for an hour at 80° C. Stir every 10 – 15 mins.
5. Leave to cool.
6. Strain the leaves from the liquid and decant this into a jar.

NEXT COMES THE MATHS

- I used 1425g of leaves in 12l of water.
- Each litre of water had approximately 118g of leaves in it. So in a 1.5l jar there would be 178g of leaves. Each 10ml of liquid has 1.18g of leaves.
- I wrote the totals on the side of the jar with a sharpie.
- Use the rhubarb-leaf mordant at 1 part liquid to 2 parts protein fibre. So 350g of wool yarn needs all the jar, and 100g yarn needs approximately 45ml of the mordant.
- Soak the yarn in the rhubarb-leaf mordant and water for several hours before dyeing in the usual way.

Wool yarn dyed with madder

When you have used all the rhubarb leaves, why not use the tasty stems too?

Try poaching them in a little water and orange juice with sugar and ginger if you like it. I like serving this with a family-favourite, ginger biscuits.

Ginger biscuits

- 100g margarine or butter
- 1 generous tbs golden syrup
- 350g self-raising flour
- 100g demerara sugar
- 100g light muscovado sugar
- 1 tsp bicarbonate of soda
- 1 tbs ground ginger
- 1 beaten egg
- 25g crystalised ginger, finely chopped

1. Preheat the oven to 160°C/325°F/Gas 3
2. Measure the margarine and golden syrup and gently heat together until all the margarine has melted.
3. Mix all the dry ingredients together in a large bowl then add the melted margarine and syrup, and the egg. Mix together well and roll small amounts (the size of a walnut) into balls. Place on greased baking trays allowing room for the biscuits to spread.
4. Bake in the pre-heated oven for 15-20 minutes until golden. Lift off the baking trays and leave to cool on a wire rack.

Gingernuts with poached rhubarb

RED CABBAGE DYE TRIAL

I have also done some experiments with red cabbage as a dye. On the internet I have seen seemingly great colours on yarn and fabrics. I had always been taught that red cabbage is a fugitive, i.e. it doesn't hold or last, but I thought I'd give it a try anyway.

HERE'S WHAT I DID

1. Chop up a small red cabbage into very small pieces.
2. Put this into a large pan with 2-3l of water.
3. Boil for 1 hour to extract the colour.
4. Strain the cabbage water dye which should be a strong pinky red.
5. Dispose of the cabbage. It won't taste very nice but it will compost.

Red-cabbage dye is interesting in the way it changes with pH. It goes bright pink with vinegar (acetic acid) which lowers the pH and goes blue with bicarbonate of soda, an alkali which raises pH..
   I used alum mordanted yarn, a Blue Faced Leicester wool DK in each of the 3 red cabbage dye baths.
- Neutral Purple red
- Acid pink
- Alkali blue

Left to right: Alkali blue, neutral purple and acidic pink

The dye colours themselves are beautiful in the pot, but as I suspected the colour is fugitive and did not fix to the yarn very much at all. I don't think a more concentrated dye solution would make a difference to the intensity of colour, or lack of it. So, I can only assume that the photos on the internet are of yarn and fabric that hasn't been rinsed and dried.

USING PH TO CHANGE THE DYE COLOURS OF MADDER

Changing the pH of substantive dyes and mordants is an interesting way to increase the colour range of a dye as well as brightening or 'saddening' the original colour.

Using my favourite dye stuff, madder (Rubia Tinctoria), with alum mordanted yarn, here's what to do:
1. Soak the madder in water for 24 hours in a large pan (that won't be used for food afterwards).
2. Gently heat the madder in water for 45 mins – 1 hour at 60° C to create the full dye colour. Then allow this to cool.
3. Mix the alum mordant with water and add to a bucket of cold water, then submerge your yarn. Leave for at least 30 minutes.
4. Lift the yarn from the mordant solution and gently put into the dye pot. Raise the temperature to 60° degrees for 45 mins. Then allow the dye pot with yarn to cool.

When the yarn is dyed, soak it in a small amount of dye with either a few tsps. of vinegar (acid) or bicarbonate of soda (alkali).

So, the acidic afterbath makes the tones more orange and the alkali afterbath makes the tones more red.

The colours from the top are madder with vinegar and madder with bicarbonate of soda.

This Month's Pattern

This month's pattern uses a DK yarn to make some crochet mittens – perfect to keep your hands warm as you work in your dye garden.

I used spirulina powder in the same way as turmeric in January to get this beautiful pale green yarn for these mittens. It's not a long-lasting dye colour but it's still pretty while it does. It fades after several washes, and after about 12 months in the light.

March Mittens.

You will need...

100g Naturally Gorgeous merino alpaca DK yarn, 3.5mm crochet hook, yarn needle to sew in ends.

Tension

Over Double Crochet, 22 st and 24 rows = 10cm

Method

Ch 16 to cast on,
Ch1, dc into BL in every st, turn
Work this for 36 rows to create the cuff rib

Fold the rib piece in half and working across the edge (with the cuff in a cylinder shape) dc into FL of front st and BL or back st then turn inside out so the seam is inside the cuff

Right hand and Left hand the same
Ch 1, evenly space 36 dc around the cuff edge and work 2 rows dc on these st,

Sl st into first st to complete every row

Row 3, ch 1 dc 17, 2dc in next st, 2dc in next st, dc 17. (38)
Row 3+4, ch 1 dc every st
Row 5, ch 1 dc 17, 2dc in next st, dc 2, 2dc in next st, dc 17. (40)
Rows 6+7, as rows 3+4
Row 8, ch 1 dc 17, 2dc in next st, dc 4, 2dc in next st, dc 17. (42)
Rows 9+10, as rows 3+4
Row 11, ch 1 dc 17, 2dc in next st, dc 6, 2dc in next st, dc 17 (44)
Rows 12+13, as rows 3+4
Row 14, ch 1 dc 17, 2dc in next st, dc 8, 2dc in next st, dc 17 (46)
Rows 15+16, as rows 3+4
Row 17, ch 1 dc 17, 2dc in next st, dc 10, 2dc in next st, dc 17 (48)
Rows 18-22, as rows 3+4
Row 23, ch 1 dc 17, 2dc in next st, dc 12, 2dc in next st, dc 17 (50)
Rows 24-28, as rows 3+4
Row 29, ch 1 dc 19, miss 12 st (of thumb), ch1 dc 19 (38
Rows 30-37, ch1, dc every st, including ch1 (from row 29) (39)

Optional frilly edge,
Ch1, dc, ch 3, dc* repeat across the row, sl st to join.

cast off.
Re-join yarn st the first st of the thumb edge,

Work 4 rows ch1 dc with sl st to join ends of rows, cast off. Repeat optional frilly edge if you wish.

Abbreviations

alt – alternate
beg – beginning
bet – between
BL – back loop
bl – block
BPdc – back post double crochet
ch(s) – chain(s)
cl(s) – clusters
dc – double crochet
dec – decrease
dtr – double treble crochet
edc – extended double crochet
ehdc – extended half double crochet
esc – extended single crochet
fdc – chainless foundation double crochet
fhdc – chainless foundation half double crochet
FL – flont loop
FPdc – front post double crochet
fsc – chainless foundation single crochet
hdc – half double crochet
hk – hook

inc – increase
lp(s) - loops
p – picot
pat(s) – pattern(s)
pm – place marker
rem – remain(ing)
rep, *, [] – repeat(s)
rnd(s) – round(s)
sc – single crochet
sk – skip
sl st – slip stitch
sp(s) – space(s)
st(s) – stitch(es)
tog – together
tks – Tunisian knit stitch
tps – Tunisian purl stitch
tr – treble crochet
tr tr – triple treble crochet
tss – Tunisian simple stitch
ws – wrong side
yo – yarn over

APRIL

April – A Tanka Poem

I like gentle April,
It is spring and blossoming.
With sprinkling showers, it lightly flowers,
April sings and I am happy.

It is not certain how April got its name. It may come from the Latin 'aperire' which means 'to open', like the flowers which bloom in April.

In Anglo-Saxon, it was called 'Eostre monath' from the godess Eostre, and the word Easter comes from this.

The weather in April is traditionally showery after the winds in March. Average temperatures range from high of 12° to low of 4°.

It's been quite cold and windy here this month. My dye plant seeds have been reluctant to germinate so I'm leaving them in the hope that they will grow when the weather warms up. Some have grown and I will be potting them on before they go into the flower borders or tubs in the dye garden.

TSome cotton fibre dyes more successfully than others because they have a mercerised (alkali) coating which has a better affinity to dyes. Cellulose/plant fibres need different and more robust mordanting because they are tougher for the dyes to penetrate.

IN THE DYE GARDEN

Now is the time to sow your calendula and coreopsis if you have chosen these for your dye garden. Plant in seed compost in trays or small pots following the instructions on the packets. Both types of seed are quite large and easy to handle, perhaps you will have a little helper with these? Children love planting seeds and helping.

When your dye-plant seedlings are large enough to handle, gently lift them in small clumps of 4 or 5 with a dinner fork. If they are large enough, separate them into single plants, or leave in their little clumps. Then either pot on by placing into small pots with compost to grow a bit bigger, or if large enough and the risk of frost has passed, plant straight into the tubs/ground where you want them to grow to full size.

You will hopefully have many more dye-plant seedlings than you need so why not do a swap with a friend? especially if they have grown different dye-plant seedlings.

MORDANTING EXPERIMENTS

After a bit of research, I have been doing some mordanting experiments with cotton yarns.

Aluminium Acetate (Al-Ac) is the usual mordant for cotton, with a boil scour first. I have compared this method to a scour with washing soda, and a three-stage mordanting process with alum and washing soda, then Tannic acid, then alum and washing soda again.

The microscopic structure of protein fibre like wool is quite different to cellulose fibres like cotton, bamboo, linen and hemp. Wool fibres are made of an open structure with a core, called the cortex, and this is covered with overlapping scales. When the cortex swells with water, the scales open up, allowing mordant and dye to penetrate. Different breeds of sheep fleece react and take dyes in varying depth depending on thickness, and scale size and overlap. Cellulose fibres like cotton are coated in a waxy substance, and if the cotton is not scoured, the wax will resist the penetration of dyes. Some cotton fibres dye more successfully than others because they have a mercerised (alkali) coating which has a better affinity to dyes.

Woad seedlings

THE COTTON SCOUR/MORDANT EXPERIMENT GOES LIKE THIS

Mark two x 100g skeins of cotton yarn with a loop of red cord on each and scour with soda ash. Carefully add 35g of soda ash to 12l of water in a large pan. (Be very careful as it can bubble up.) The cotton yarn skeins are added to the pan and stirred with a long spoon. Bring the pan to the boil then simmer for 2 hours stirring gently now and again. Allow the pan to cool and pour the dirty water off. The process can be repeated if the water is very dirty.

1. Place two more 100g skeins of cotton yarn (unmarked) in plain water and boil for 2 hours. Allow to cool. Pour the water away, the process can be repeated if the water is very dirty.

2. Take one skein from each of the scouring processes and add a blue loop to each.

3. The blue loop skeins are both mordanted using the 3-stage process with alum.

4. Dissolve 50g alum, 12g washing soda (for 200g cotton yarn) in hot water in the pan then add the cotton yarn to the pan with more hot water. Bring the pan to a simmer then turn off and cool overnight. Remove the cotton-yarn skeins and dispose of the alum/washing soda solution down the drain.

5. Dissolve 12g tannic acid in hot water in a pan. Add the cotton yarns (marked with a blue loop) and top up with enough hot water to cover. Simmer for 30 mins then leave to cool overnight. Remove cotton yarns, dispose of solutions then repeat step 1 with alum and washing soda again.

6. Rinse the cotton yarn skeins thoroughly. They can now be dyed, or dried and stored for use later.

7. Take the remaining two skeins (one with a red loop only and one without any loops) and soak in warm water overnight in a pan. Dissolve 10g Aluminium Acetate (Al-Ac) in a jug with hot water then add to the pan. Simmer for 1 hour then allow to cool. Remove the yarn and rinse well. It is ready to dye. If you wish to store the prepared yarn until later, dry it without rinsing off the Al-Ac. Then rinse just before dyeing.

From bottom left to top right, boil scour & Al-ac mordant,(no loops) soda ash scour & Al-ac mordant, (red loop only) soda ash scour & 3 stage mordant,(red loop and blue loop) boil scour & 3 stage mordant. (blue loop only)

Personally, I prefer to use alum mordant as it is more commonly available and user friendly. The best method for colour saturation and richness was the soda ash scour with 3 stage method mordanting. The least satisfactory method was the boil scour with Al-Ac mordant. What do you think?

All these skeins were dyed together in a Logwood (Haematoxylum Campechianum) dye pot. I chose this dye because it gives a strong colour that dyes cotton well.

Cotton fabrics were also scoured and mordanted in the same ways and the results were the same, with the Soda Ash Scour and 3 stage mordanting being the best colour result.

A BIT ABOUT ROOIBOS

Rooibos (pronounced Roy-boos) is a herb which grows in the Cedarberg mountains in South Africa. The plant likes the dry mountainous climate with periods of extensive rainfall.

Rooibos, *Aspalathus Linearis* is a member of the legume family of plants which also includes peas, beans and chick peas. It has long thin leaves which give the tea, and it has seed pods like peas and beans with a single seed inside. Seeds take 18 months to grow from seedling to harvest-ready plant and are usually harvested in the South African summer months.

The leaves are harvested and then oxidised to release the rich red colour and sweet, malty flavour. It is most often drunk as a tea, without milk, and sometimes honey is added for extra sweetness. The flavour of Rooibos is described as sweet, honey, vanilla and caramel.

DYEING WITH ROOIBOS

Still using store-cupboard items to try to dye with, I used up some old Rooibos/Red bush tea bags.

I used alum mordanted 100g skein of wool yarn. I pre-soaked 10 teabags in boiling water in the dye pan and left to cool overnight.

The yarn was then immersed in the dye pot tea dye and simmered at 60° for 30 mins and left to cool. Then rinsed and dried as usual.

The Rooibos tea gave the yarn a soft honey beige colour. Rooibos is rich in flavonoids including luteolin which is the main colour substance in weld. (yellow). It's not a statement colour but will be a nice neutral in contrast to a bolder colour like madder or cochineal in colour work, Fair Isle, intarsia etc. It also looks lovely on its own in this Delta shawl pattern.

Rooibos dyed yarn and delta shawl detail

Here's the Delta Shawl Pattern

Project description

This shawl is for experienced knitters who can read their knitting and anticipate where the patterning stitches should be. Ideal to use that 100g of hand dyed 4ply yarn you just couldn't resist!

You will need...

100g Naturally Gorgeous 4ply hand dyed yarn, or Purely Gorgeous 4ply
3.5 long needles or long circular needle, yarn needle to sew in ends, stitch marker

Tension

23 st and 32 rows over the Delta points after blocking. Tension isn't vital, but should not be tight.

Special stitches & details

The eyelets are created by YO, K2tog (RS) or YO, P2tog (WS) on the first half of each row and come together to form a Delta pattern and contribute to the shaping.

Method

CO 20st and work 2 rows stocking st. PM after 10st, this will mark the 'midline' of your shawl.

RS. Row 3. CO 5st, put back on left needle and work as follows K4 YO K2togbl, K to end, SM as you go.
WS. Row 4. CO5, put back on left needle, P4 YO P2tog, purl to end SM as you go.

RS. Row 5. CO 5, put back on left needle, K4 YO K2togbl, K4 YO K2togbl, K to end, SM as you go
WS. Row6. CO5, put back on left needle, P4 YO P2tog, P4, yo, P2tog, P to end, SM as you go

RS. Row 7. CO5, put back on left needle, (K4 YO K2togbl)* rep * 3 times, K to end, SM as you go
WS. Row 8 CO 5, put back on left needle, (P4, YO, P2tog,)* rep* 3 times, P to end SM as you go.

Keep repeating this 2 row pattern, increasing 5 stitches and a (repeat)* on each row. The eyelet holes will move towards the midline marker creating the Delta shape.
When the eyelet rows meet in the middle, work in the same way with the (repeat) pattern but on RS row to work the last eyelet repeat you will cross the midline. Remove the marker temporarily.
Continue working in this way, moving eyelet holes to the midline and increasing by 5 st each row until you have 280 st, then on the next row:

RS. CO 5 st and move to left needle, then work these first 5 st in Moss/Seed stitch, K1,P1,K1,P1, K1 then work the eyelet (repeat)s to the midline as before and then K to end of row.
WS. CO5st and move to left needle, work in Moss/Seed st for 5 stitches, K1,P1,K1,P1,K1 then (repeat)s to midline and purl to last 5 st, K1,P1 etc Moss/Seed st.

RS. Increase 5 stitches in the same way, work Moss/Seed st for 10 st, starting P1,K1 etc, then work (repeat)s to midline then K to last 5 st, K1,P1 Moss/Seed st.

WS. Increase by 5 st, as before, work Moss/Seed st for 10 st, starting P1,K1, then (repeats) to midline, then P to last 10 st and work Moss/Seed st, P1,K1 etc

Continue increasing by 5 st each row, and work Moss/Seed st as they present, then (repeat)s etc until you have increased by 20st each end of the shawl, 320 st total.

Work 20 rows of Moss/Seed st, eyelet (repeat)s etc without increasing, but moving the eyelets as before and taking the st at each end into the Moss/Seed st pattern.

Work 5 complete rows of Moss/Seed st across the whole row without any eyelet (repeat)s then BO loosely.

Block your work to open up the Delta patterning to full effect.

Delta shawl stitch detail

Abbreviations

K – knit
P – purl
st(s) – stitch(es)
m – metres
yds – yards
cm – centimetres
in – inches
mm – millimetres
g – grams
LH – left hand
RS – right side
WS – wrong side
Rep – repeat
MC – main colour
CC – contrast colour
PM – place marker
SM – slip marker

inc – increase as described in pattern
dec – decrease, usually by knitting of purling two together
Sl1 – slip one st (purlwise unless directed otherwise)
Yo – yarn over (also known as yarn forward or yarn round needle)
Kfb – knit into front and back of next stitch
K2tog – knit two together
K3tog – knit three together
K2togtbl – knit two together through back of loops
K1tbl – knit one through back of loop
cdd – centred double decrease – slip 2 sts knitwise together, knit 1 stitch, then pass 2 slip sts over the knit stitch
puk – pick up and knit into front of loop lying between stitch previously knitted and following stitch
sl1wyib – slip 1 with yarn in back
sk2p – sl1, k2tog, psso
Sl1 making ds – after turning the work, slip the first stitch purlwise with yarn in front. Then take the yarn over the needle, pulling the stitch from the previous row over the needle to form a double stitch. Work next stitch as instructed. When you reach a double stitch in a following row, simply treat as 1 stitch and knit the 2 strands together.

I like making this South African Rooibos Shortbread recipe that is really delicious with a cup of tea, Rooibos or otherwise.

Rooibos Shortbread

- 1 cup unsalted butter (I use Goat's butter because I like the taste)
- 2 tbs/5 tea bags of Rooibos tea (I used honey and fig flavoured)
- ½ cup Caster sugar
- 2 cups plain flour

For the lemon glaze:

- 1 cup icing sugar
- 2 tbs lemon juice

1. Cream together butter and sugar until soft and fluffy. Add dry tea and leave for 1 hour for the flavour to develop. Add the flour; initially it will be quite crumbly but keep mixing and the dough will form into a ball. Roll out to approximately ½ cm thick on a floured board. Cut out the dough into shapes that you like. Gather up the remaining dough and roll out again. Repeat until all the dough is used up.
2. Place the cut dough on 2 baking sheets covered with greaseproof paper.
3. Bake at 180°C/350°F/Gas 4 for 15-20 minutes until golden brown. Lift carefully from the baking tray and cool on a wire rack. Mix the icing sugar with the lemon juice to make a smooth paste then drizzle over the biscuits.

MAY

Poem for May

Pollen in May flowers,
the bumblebee buzzes
A wild fuzzy dance

May is named after the Greek goddess Maia. It was first called May in about 1430.

The Anglo-Saxons called this month 'Tri-milchi' to recognise that the lush grass meant cows could be milked three times a day.

In most parts of the UK May is celebrated as the end of winter and the beginning of summer. 1st May was celebrated with dancing on the village green, often winding ribbons around a May pole and with traditional folk/morris dancing.

The weather temperatures range from high 16° to low 7° with about 10 days of rain.

IN THE DYE GARDEN

At last, the weather has warmed up enough to plant out the seedlings into the tubs in the dye garden. The woad, chamomile and safflower seeds I planted are all doing well. The Japanese Indigo, golden rod and dyer's greenweed have not germinated at all and I've given up on them.

The marigolds *(Calendula)* and coreopsis seeds were planted at the end of last month and are sprouting. I enjoy using calendula as a dye, either fresh or dried. It gives a beautiful rich golden yellow that works well with so many other natural dye colours. It's not particularly light-fast, but I enjoy the colour anyway while it lasts.

Here are the woad, tansy, chamomile and safflower plants in the tubs in the dye garden:

This month the gorse flowers around us in Cornwall seem more prolific than usual. Gorse tends to flower all year round but in the late spring and summer months it's at its best. This is because there are at least two species of gorse which grow in Cornwall; the Common gorse, (Ulex europaeus) and the western gorse (Ulex gallii) . The first flowers January to June and the other from July to November. Hence the Cornish song, "kissing's out of fashion when the furze is out of bloom"!

Gorse flowers can give beautiful yellow colours, from soft primrose-yellow to a warm lemon-yellow.

Left to right: Woad, tansy, safflower and chamomile, madder in the middle of the raised bed

GORSE (ULEX EUROPAEUS)

Gorse is closely related to Broom and is a relative of the Pea family. It thrives in dry, sandy conditions, especially heath and scrubland. The shoots and leaves of gorse have adapted and become very sharp, hard thorns. All gorse has yellow flowers and generally a long flowering season. Hence the Cornish song, "kissing's out of fashion when the furze is out of bloom"! Gorse flowers have a subtle coconut smell and are edible but the pea-like seed pods are TOXIC. Gorse had been used throughout history as a barrier to keep stock in or out of an area and it provides shelter for many insects and birds, including the Yellowhammer. Other uses include as fuel for bread ovens because it burns very hot, as fodder for animals and its branches bound together can be used as a broom on floors or up chimneys.

Gorse was traditionally added to a bride's bouquet to symbolise fertility. In Wales it was burnt to deter witches and it was used for the May fires of Beltane in Celtic traditions.

DYEING WITH GORSE FLOWERS

The flowers can be used fresh to create a dye bath at about 2:1 WOF i.e., 200g flowers to 100g dry yarn or fibre. Gorse dyes protein fibres like wool, alpaca and silk best.

CAREFULLY pick the gorse flowers with thick gloves on. The thorns on gorse are tough and sharp and can really hurt your hands. It will take a while to pick enough, but it's

Gorse flower dye

worth it to enjoy the lovely soft coconut smell the flowers give off as you pick. Pick extra for the gorse flower cordial recipe too.

TO PREPARE THE DYE

Tip the gathered gorse flowers into a large pan as soon after picking as you can. Add 2-3l of water and bring to a simmer for about an hour so the flowers give all their colour. Then either strain off the flowers from the dye liquid or use it as it is with the flowers in the liquid. You will have to pick flowery bits out of your yarn or fibre, but some dyers feel it gives a richer colour to dye 'all-in-one' like this.

Put the alum mordanted yarn into the dye pot and bring to a simmer at 60-80°C for 45 minutes. Allow to cool completely. When cold, lift the yarn/fibre from the dye pot and rinse in lukewarm water (picking out any stray bits of flowers). Keep rinsing until no more colour shows in the rinsing water. Dry your yarn gently away from direct heat. Your yarn is now ready to use.

ABOUT DYE CONCENTRATIONS

Gorse gives quite a soft, pale primrose yellow. It is tempting to think that by using more flowers or dye stuff you will get a stronger, darker colour. However, this isn't the case. The amount of dye colour molecules in the dye pot doesn't alter the amount of colour molecules the fibre can absorb. So, if your dye pot has oodles of molecules and the fibre can only absorb a maximum of half of them, that's its maximum or darkest colour. The remaining dye can be used again though because there are still colour molecules available to dye with. The dye can be used again to 'Exhaust' dye for paler tones. (This will be explored next month.)

The amount of water in a dye pot doesn't affect the strength of colour either. If you have a dilute solution – i.e. the maximum colour molecules (for the right amount of dye for weight of fibre) in a lot of water – it will appear more dilute and clearer, but will usually give the same depth of colour as the same amount of dye in less water which appears to be a stronger solution.

Gorse Flower Cordial Recipe

While you are picking gorse flowers try to pick more than you need to dye with so you can make this delicious spring cordial.

- 4 handfuls of Gorse flowers
- 600ml cold water
- 250g caster sugar
- Zest of an orange and juice of a lemon

1. Bring the water and sugar to a rapid boil and keep boiling for 10 minutes. Remove the pan from the heat.
2. Grate the orange and juice the lemon.
3. Add the zest, juice and gorse flowers to the sugar water. Stir well and leave until cool or overnight.
4. Strain the liquid through a muslin or fine-mesh sieve into a clean container like a glass jug.
5. Pour into a sterilised bottle, cap and store. Refrigerate once opened.
6. Serve with sparkling water or champagne!

Gorse Blossom Cordial and champagne, the perfect May drink

Gorse Blossom Cardigan

Teddy is wearing her Gorse Blossom Cardigan

This Month's Pattern is for a child's cardigan, knitted in gorse dyed DK.
A pretty long-sleeved, round neck cardigan for a little girl, knitted in hand dyed DK or 4ply yarn
Instructions for DK are in **Black** and 4ply are in Blue if they differ from the DK version.

You will need...

200g of DK yarn or 100g 4ply. I used hand dyed merino/alpaca for a super cosy knit.
4.00mm needles or 3.5mm needles, yarn sewing needle, 6 or 5 x 1cm buttons

Tension

21 st x 32 rows = 10cm on 4.0 mm needles over stocking stitch. Finished chest size 25cm/10Inches
27 st x 36 rows = 10cm on 3.5mm needles over stocking stitch. Finished chest size 21cm/8.25Inches

Special stitches & details

To create the eyelet holes, YO K2tog

Method

For the back
Using 4.00mm/ needles, C/O 65 st.
Work 6 rows of garter stitch (Knit every stitch, every row)
RS Row 7 K
WS Row 8, every following WS row, Purl
RS Row 9 create eyelet row: K5, YO, K2 tog to last 4 st, K4, every following RS row, K
WS Row 10, Purl
RS Row 11 K2 K2togbl to last 4 st, K2tog, K2

Continuing to work in Stocking st, Repeat decreases every 10th following rows until 55 st (Approx 20cm/8' 17cm/6.75inches)

To shape the armholes, cast off 4 st at beg of next 2 rows. (47 st)

Continue in stocking stitch without shaping until armhole measures 12cm/4.5' 9cm/3.5inches ending with a WS row

To shape the shoulders,
Cast off 6 st at beg of next 4 rows (23 st)
Place remaining 23 st at the back of the neck on a stitch holder.

For Left Front
Using 4mm 3.5mm needles, C/O 32 st.
Work 6 rows of garter stitch.
RS Row 7 K
WS row 8 P, every following WS row, P
RS Row 9 create eyelet row: K5, YO, K2 tog to last 4 st, K4, every following RS row, K
WS Row 10, Purl
RS Row 11 and very following row, K to last 3 st, YO, K2togbl (Centre Front edge)

At the same time, decrease on every 10th row on the side seam edge by K2 K2togbl

Continue until you have 26 st, and work measures the same as the back to the armhole.

To shape the armhole,
cast off 4 st at side edge of next row (22st)

Continue in stocking stitch until the armhole measures 6cm/2.5inches 4cm/1.5Inches ending with a RS row.

To shape the front neck
Cast off 5 st (neck edge), p to end (17st)
Work 9 rows, decrease 1 st at neck edge next and every following alternate row (12st)

Continue without shaping until armhole measures 12cm/4.5' ending with WS row

To shape the shoulder
Cast off 6 st, knit to end
Next row, P
Cast off remaining 6 st.

For Right Front
Repeat as for Left front reversing shaping, and *replacing* YO, K2togbl at front edge *with* K2, YO K2togbl.

Sleeves, Both alike
Using 4mm 3.5mm needles, C/O 37 st.
Work 6 rows garter stitch

Increase 1 st at each end of next and every following 6th row to 45 st.
Continue without further shaping until sleeve measures 19cm/7Inches' 17cm/6.5inchesending with a WS row.

To shape sleeve top
Cast off 4 stitches at beg of next 2 rows (37st)
Cast off 1 at beg and end of next 6 RS rows (19st)
Cast off 5 st beginning of next 2 rows, (9 st)
Cast off remaining 9 st..

Neck band and button band/front edging
Join shoulder seams, and with RS facing, using 4mm needles, pick up and knit 23 st from right side of front neck, 23 st from stitch holder and 23 st from left front neck, ending at left centre front edge. Knit 6 rows garter stitch and loosely cast off.

With RS Right front facing, using 4mm 3.5mm needles, pick up and knit 3 st from garter stitch edge, 47/44 st along front edge and 3 st from front neck garter st edge.
Work 3 rows garter st starting with WS row (Neck edge)
Row 4 K3 (K2tog YO K8)* rep 6 times, K2
Row 4 K2 (K2tog YO K6)* rep 5 times, K2
Work 2 further rows garter st, taking in YO loops as stitches.
Cast off.

Work left front band in the same way, working 6 rows of garter st and omitting the button hole row.

Sew the shoulder seams to the body, then sew the side body and underarm seams.
Sew on 6 x 1cm buttons onto the left front band, lining up with the buttonholes.

Gorse Blossom Cardigan

Abbreviations

inc – increase as described in pattern
dec – decrease, usually by knitting or purling two together
Sl1 – slip one st (purlwise unless directed otherwise)
Yo – yarn over (also known as yarn forward or yarn round needle)
Kfb – knit into front and back of next stitch
K2tog – knit two together.

JUNE

Poem for June

May blossoming June
Flourish in greening sunshine
Plants smiling

June is named after the Roman goddess Juno, the Goddess of Marriage. This is why June is considered to be the best month to marry. 'Married in the month of roses, June, life will be one long honeymoon.' The flower of June is the red rose, also associated with love and marriage.

The longest day of the year is 21st June when the sun is at its most northerly point and the hours of daylight are longest. The sun is also at its zenith, or highest in the sky in June so temperatures are a high of 19° and low of 10°. Rainfall averages 62mm.

Midsummer's day was a magical time associated with fairies, witches and dancing.

IN THE DYE GARDEN

My dye garden has flourished with all the sunshine (and a bit of rain) we have had. The chamomile and tansy are almost in flower- I'm sure that by the end of the month I will be picking a steady crop to dry and store.

A friend gave me some amaranth plants to grow on. They give a red dye that I've never used before so that will be new learning for me. There's not much information available on how to use it to dye

Amaranth

with so it will be an experiment. The amaranth plant is edible; the seeds can be eaten raw or sprouted, the leaves and stems can be used like spinach.

I have a few nettles in the garden which are persistent and tough. Rather than resort to weedkiller I have cut them back and I'm going to use them to dye with. I don't have enough, so I enjoyed a lovely walk last week with my amaranth-giving friend, gloves, secateurs and trug ready to cut nettle tops for the dye bath.

EXHAUST DYEING

The dye bath should have some colour left so you can try to exhaust dye. This uses the principle that colour molecules left in the dye bath after the first dyeing can be used to dye a second batch of fibre or yarn. Sometimes a dye bath will give a second and third exhaust dye. Madder and cochineal are especially good for this. I found that the nettle exhaust dye gave very little colour.

To exhaust dye, simply use the remaining dye with more mordanted yarn in the same way as you did the first-time round. The colour will be paler.

COLOUR ON DIFFERENT YARN BASES

The colour intensity of dyes can change slightly according to what fibre you are dyeing onto. For example, merino and silk take colours more strongly than cotton and linen. Within wool yarns from different breeds of sheep there can also be subtle differences too. Within wool yarns from different breeds of sheep there can also be subtle differences too. This may be due to the different scale pattern of the fibre which absorb dyes differently as well as the reflection of the colour back at you. So, some finer wools with smaller, less prominent scales absorb the dye well, and some of the lustre longwools reflect more light back at you, making the colour appear richer.

In my experiments, the yarns were mordanted together, dyed together and gave subtle colour variations in a chamomile dye. (All yarn samples from West Yorkshire Spinners)

As you can see, the Shetland Island wool took up the colour from the chamomile most strongly and the Blue Faced x Kerry Hill wool took it least strongly. If you are dyeing for a large project then it is advisable to use the same yarn type throughout, and to dye as much as you can in the same dye pot at one time. Unless of course you want to introduce the different textures and shades of different yarn types to your project.

Colour on different yarn bases; top to bottom, Blue Faced Kerry, Blue Faced Leicester, Falkland wool with silk, Falkland wool with alpaca, British wool, Shetland Island wool.

STINGING NETTLES

Stinging Nettles *(Urtica doica)* are found from spring to autumn in most unkempt and wild land in the northern hemisphere. Nettles have been used in many ways through the ages, including as a spring vegetable or as a tonic pick-me-up after the winter. The leaves are used to wrap cheeses, as a tea, even a spicy beer. Nettles are rich in iron and vitamin C. Cooking them destroys the stinging hairs and makes them safe to eat.

Nettle stalks were traditionally used as a fibre, a bit like flax. The stem outer layers were stripped (Including the stinging hairs), the tough fibre retted and the softened inner fibres then carded and spun to produce a string cord or rope. The word 'ret' is a corruption of the word rot. The stems are laid on damp ground, or in flowing or stagnant water. The bacteria that develop attacks the pectic that holds the fibre, core and bark together and softens them.

The European Stinging Nettle is difficult to grow commercially so most nettle fibre is from the Himalayan Nettle (Girardinia diversifolia). As a fibre source, nettle is much more sustainable than cotton because it needs less water to grow and process and needs few if any pesticides. If the nettle fibre is to be used as a yarn it is usually blended with wool. Without it, the nettle fibre lacks bounce and flexibility. Nettle fibres are smooth and silky.

Make sure you use gloves!

Stinging-nettle tops

There are a few commercial nettle yarn producers who use nettles in their wool blend – the mix is generally 70% wool, 30% nettle fibre. A fine yarn with the addition of nettles is ideal for socks because the nettle fibre gives the finished socks strength and durability.

TO DYE WITH STINGING NETTLES

To prepare a dye bath with stinging nettles you need to do the following:

Pick about 300 – 400g fresh nettles/100g of yarn or fibre. Then pull off the leaves and chop them up with scissors. You need to end up with about 200g - 300g chopped leaves/100g yarn or fibre.

Put the leaves in your dye pot and pour over boiling water so the leaves are soaked and floating. Leave this to infuse for at least 24 hours. Then simmer the nettles for 30 – 40 minutes. Allow this to cool and strain it into another pot ready to use. The nettles can be composted or used as mulch. (They are really good for the garden.)

Use an alum mordant (as described previously) and soak your yarn/fibre. I used a wool/nettle mix yarn. The wool content is 70% and the nettle is 30%. Therefore, as the protein fibre (i.e. wool) content is higher than the cellulose content (nettle) it is best to mordant with alum. I haven't tried nettle dye on cellulose fibres. I suspect the colour might be poor as it is difficult to get plant dyes to fix to cellulose fibres with any real colour depth.

Nettle dyed yarn

Place your mordanted fibre into the cool dye bath and gently simmer for about 45 mins. Allow to cool then rinse the yarn until the water is clear and colourless. Nettle dye should give a soft green-brown, like khaki. The early summer plant tops give a better colour than the tougher late summer ones so June is a great time to experiment. Nettles have been a food source for hundreds of years – most usually as soup, and they have a strong, spinach type flavour.

Here is a recipe for this vibrant green soup if you fancy trying it. A bit like watercress soup but with a more earthy taste.

Nettle, Pea and Watercress soup

Soup in the shade in the garden

Ingredients:
- 1 tbs olive oil
- 200g frozen peas
- 1 medium white onion
- 200g fresh nettle tops
- 1 clove garlic
- 200g watercress
- 800ml vegetable stock
- 3 tbs green pesto
- Wild garlic flowers to garnish

Method:
1. Use the small bright-green tender nettle tops for this soup.
2. Heat oil in a large pan and sauté onions and garlic for 3 minutes until soft. Add the stock and frozen peas and cook gently for 2 minutes. Add the nettle tops and watercress and boil for 3 minutes.
3. Remove from the heat and blend until smooth. Stir in the green pesto and add a swirl of cream and a sprinkle of wild garlic flowers. Serve with bread and cheese, preferably Yarg cheese which is wrapped in nettles!

This month's pattern is for socks, knitted in 30% nettle fibre yarn and dyed with nettles.

Nettle Socks

By Caroline Bawn / Gorgeous Yarns

Project description

This sock pattern is knitted with a fine 4ply, hand dyed with nettles. The design features a lacy pattern which runs down the front of the leg and foot which itself has a leaf pattern reminiscent of nettles. The pattern is one size for the calf measurement. The length of the foot can be adjusted to suit you.

You will need...

150g 4ply hand dyed yarn, or Purely Gorgeous 4ply
2.25 - 3.5mm long double pointed needles or long circular needle, yarn needle to sew in ends, stitch markers

Tension

32 st and 45 rows over stocking stitch after blocking. Tension isn't vital, but should not be tight.

Special stitches & details

This is a charted lace pattern, suitable for confident knitters.

Method

C/O 64 st (in CC if using) and arrange over 4 needles. PM to mark the start of the row. The stitches should be arranged with 16 on each needle, with the marker at the centre of one needle. This will ensure your lace pattern stitches are all worked on 1 needle. Work in the round.
Work 20 rounds of K2, P2 rib, slipping the marker each time you come to it.
Change to MC and knit 24 st, make 1 st by puk, then commence lace panel as chart over the new stitch and next 16 st.
Work the remaining stitches of the round as K to the marker. Work round 2 and all even number rounds as K
Work the lace panel on the 17 stitches on odd rounds as charted.
Continue until you have completed 1 whole lace pattern panel repeat, and then to row 30 on second repeat.

Heel Flap.
Knit 15 st from marker, turn
Row 1: sl first st, P 29 st, turn

Row 2: sl first st, K 29 st, turn
Work back and forth on these 30 st for 30 rows total.

Shape the heel.
Row 1: sl first st, P16, P2tog, P1, turn
Row 2: sl first st, K5, K2togtbl, K1, turn
Row 3: sl first st, P6, P2tog, P1, turn,
Row 4: sl first st, K7, K2togtbl, K1, turn,
Continue in this way taking in an extra st every row until all the st from the heel flap are included. You will then have 18 st on your needle. Remove the marker at the back of your work which marks the start/finish of a round.

Gusset.
Join in the MC .
Pick up 15 st knitwise down the side of the heel flap, PM,
knit stitches on needles to the base of next heel flap, working the correct lace panel row as you go, pm
Pick up 15 st knitwise up the side of the heel flap and bring all the stitches back into the work.

Gusset shaping.
Round 1: K to within 3 st of the first marker, K2tog, K1, Slip marker, work stitches as set including the lace panel, to the next marker,
Slip marker, K1, K2togtbl, knit to end of the round
Round 2, K to first marker, slip marker, work lace panel as set, knit to next marker,
Slip marker and complete the round.
Repeat rounds 1 and 2 until you have 65 st, ending with a round 2. You should have 32 st at the back of the sock, and 33 st at the front of the sock.

Continue working in the round, slipping markers and working the lace panel rows as set, until your sock is 4cm shorter than your foot length when measured from the back edge of the to the tip of your longest toe. (I have UK size 5/39 feet and I completed 2 panels and 40 rows in total over the whole sock to fit my feet.)

Toe shaping.
Round 1, Join in CC if using, and work to 3 st before the first marker, K2tog, K1, Sl Marker, K1, K2togtbl, K to 3 st before next marker, K2tog, K1, Sl Marker, K1, K2togtbl,, K to end of round

Round 2, K

Repeat rounds until you have 29 st left. There will be 14 on the back of the sock and 15 st on the front of the sock. On the last round 2, after slipping first marker, K 7, K2tog, K round to last stitch before first marker again.

Arrange your stiches on 2 needles and using either Kitchener st, or a 3 needle bind off, cast off your socks.

Sew in all the ends of yarn.

Block gently and enjoy wearing your socks.

Abbreviations

K – knit
P – purl
st(s) – stitch(es)
m – metres
yds – yards
cm – centimetres
in – inches
mm – millimetres
g – grams
LH – left hand
RS – right side
WS – wrong side
Rep – repeat
MC – main colour
CC – contrast colour
PM – place marker
SM – slip marker

inc – increase as described in pattern
dec – decrease, usually by knitting of purling two together
Sl1 – slip one st (purlwise unless directed otherwise)
Yo – yarn over (also known as yarn forward or yarn round needle)
Kfb – knit into front and back of next stitch
K2tog – knit two together
K3tog – knit three together
K2togtbl – knit two together through back of loops
K1tbl – knit one through back of loop
cdd – centred double decrease – slip 2 sts knitwise together, knit 1 stitch, then pass 2 slip sts over the knit stitch
puk – pick up and knit into front of loop lying between stitch previously knitted and following stitch
sl1wyib – slip 1 with yarn in back
sk2p – sl1, k2tog, psso
Sl1 making ds – after turning the work, slip the first stitch purlwise with yarn in front. Then take the yarn over the needle, pulling the stitch from the previous row over the needle to form a double stitch. Work next stitch as instructed. When you reach a double stitch in a following row, simply treat as 1 stitch and knit the 2 strands together.

Nettle socks lace chart

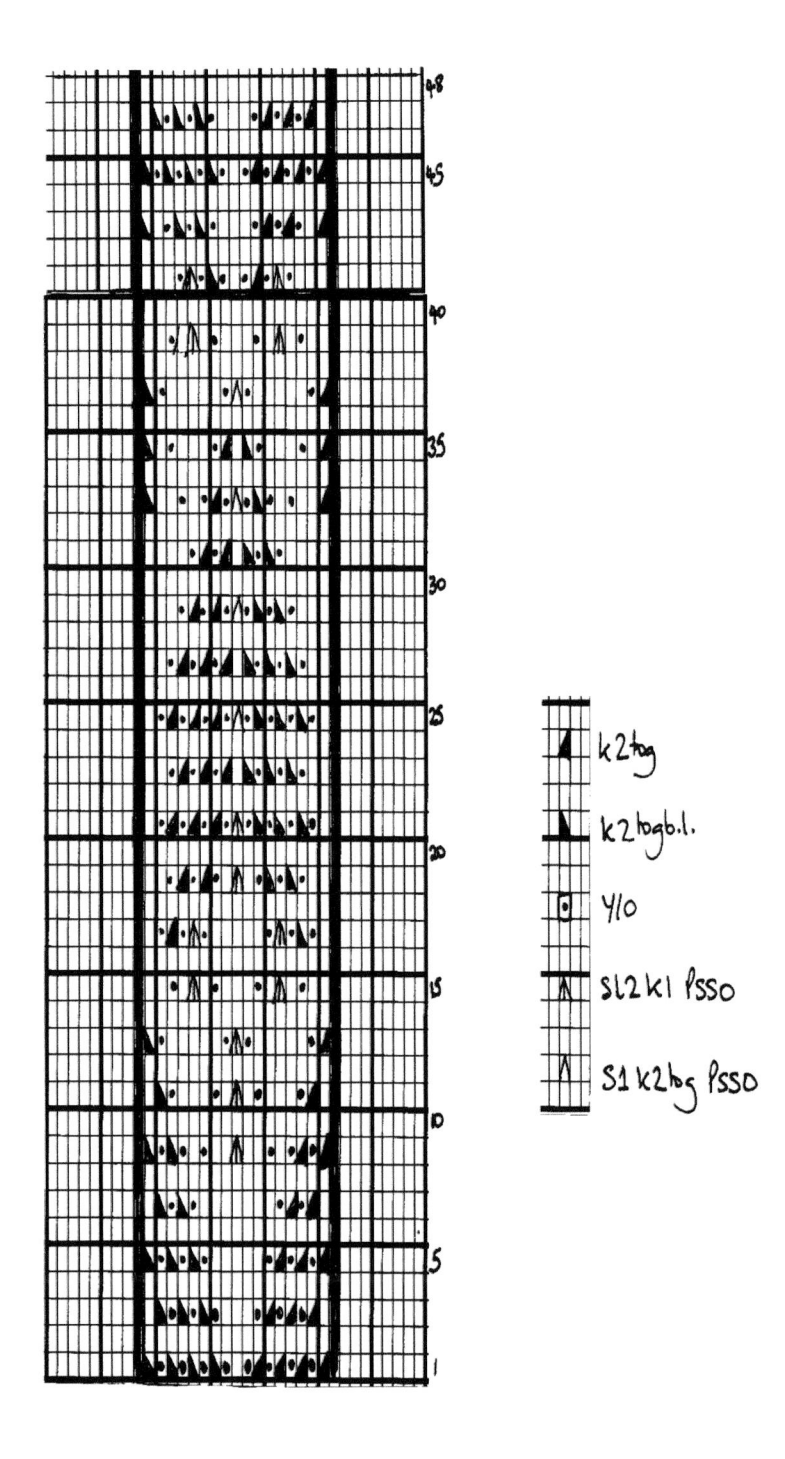

JULY

Poem for July

Sunshiny July
Tall pink foxgloves grow
Enjoying the light

July is usually the hottest month of the year with average temperatures of 21° high and 10° low and only 63mm of rainfall. St Swithin's Day is 15th July. Traditionally if it rains on this day, it will rain for 40 days after. Fortunately, the opposite is true too, so if it is fair on 15th July, there will be good weather for 40 days to come.

July is the seventh month in the Gregorian Calendar, but was the fifth month in the Roman calendar. The Romans called it 'Quintilius' which means 'fifth'. The Roman Senate renamed it 'July' after Julius Caesar who was born on 12th Quintilius.

Anglo-Saxons called this month 'Hey Monath' meaning 'hay-making month'.

IN THE DYE GARDEN

Most days this month, each morning I have been pottering in the dye garden picking the best dye flowers ready to be dried. The calendula plants have been amazing; the yellow ones have been most prolific, with the orange and orange-bronze ones flowering a bit later in the month. They all give a bright rich golden-yellow dye. This yellow can be a good base to over-dye with blue to get greens.

I use a food dehydrator to dry the dye flowers for later use, though they can be used fresh with similar colour results. If you haven't got a dehydrator, a few flowers at a time can be dried on kitchen paper on a sunny windowsill and stored for later.

As this month has been so sunny, and several weeks have been scorching hot, it's the ideal time to start a few solar dye jars. They need daylight (UV light) rather than direct, hot sunlight and take at least 3 months to 'mature' the colours. They can be left longer – up to 18 months if you forget them, as I have done sometimes. The yarn didn't come to any harm at all and the colours were really intense.

SOLAR DYEING - HERE'S HOW YOU DO IT

Use a 1.5l Kilner type jar with a wire clip top lid. (These are designed to take a bit of pressure which may build up if your dye jar gets hot or ferments.)

If you use dried dye stuff, then between 50g – 100g/100g yarn in total is enough. If you are using fresh dye stuff its between 100g – 200g/100g yarn. Fresh dye stuff tends to ferment and/or stagnate more than dried so just be aware, your yarn may be VERY smelly when you take it out.

Layer a skein of dry wool alternately with dry dye stuff in the jar. Try using madder/chamomile/cochineal in layers for a sunset palette. You don't have to be accurate, just try not to let the dye stuff get too mixed up.

Mix the alum and Cream of Tartar mordant with water and pour into the jar. The dye stuff will rehydrate and the yarn will absorb the mordant liquid.

You will need to top up the jar a bit, ideally to cover the yarn or to the 'neck' of the jar. Close the lid and leave in the daylight for about 3 months. Don't leave in direct sunlight, a greenhouse or a conservatory where the jar might get too hot and explode.

I have mine lined up on the back doorstep on the north-facing side of the house.

Leave the jar and don't shake it. You can start solar jars at any time of year but if I start some in the winter, with less daylight hours, I leave them about 6 months at least.

N.B. if you are using cochineal or logwood chips, I strongly recommend that you put them in a gauze bag tightly tied, before putting into the jar. This is because they very easily get entangled in the fibres of the wool and are very difficult or impossible to get out!

Leave the jar, with its lid closed. If the jar looks like it's fermenting, i.e. has lots of tiny bubbles at the surface, very carefully open the jar to let out the carbon dioxide from the fermentation process. Leave it open a day or two then close it again. If your jar gets a mouldy crust, don't worry. You can carefully scoop it off with a spoon, avoiding any dust that comes off the mould. Alternatively, leave it until your

Solar dye jars

Solar Dye jar layers

yarn is ready to rinse and it will come off then. It doesn't usually damage the yarn though it may darken it a bit.

3 MONTHS HAVE PASSED

Use a large strainer or colander in the sink or over a bucket. Tip the contents of the jar into the strainer to catch the bits and stop your sink getting blocked up. Slowly and patiently rinse the yarn to get all the remaining bits out and remove any residual colour that hasn't been absorbed by the yarn. Every skein of yarn dyed in this way will be slightly different. That's the beauty of this method, you can't get it wrong!

The used dye bits can be used again with the same method, this time to create an overall mixed colour rather than the colour separation you had the first time. Any used remaining dye stuff and any rinsing water in the bucket can be safely put on the garden.

This method of dyeing is ideal for experimentation and variation. The combination of dye stuff in layers creates a multi-tonal, multicolour effect with extra colours where the dyes meet. The usual colour mixing rules apply. A yellow layer (chamomile) meeting a layer of cochineal (dark pink) will give orange tones.

The Results!

FORAGED DYE THIS MONTH

Dock plants in the garden

Following last month's foraged dye stuff, nettles, this month a new weed has taken its place as top nuisance weed. It's dock, common names include bitter or broadleaved dock. . (Rumex obtusifolius) It's Latin name makes me smile. It certainly is obtuse in where and how it grows, and its stubbornness in removal. It grows all over and is easily foraged for the dye pot. I have never dyed with dock before but I thought I would give it a go. Other dyers have got yellowy, honey beige from the leaves and rich brown from the roots. I have read that older plants give darker colours and fresh spring leaves give a vibrant yellow.

THIS IS WHAT I DID

I gathered about 400g of dock leaves and 300g root from one large plant in the garden. Dock has a very long, thick root that can be deep and tough to dig up. Like many weeds, it will regrow from any bits of root that are left. That's the obtuse bit showing itself again!

The leaves were roughly cut up with scissors and simmered gently for about 40 – 50 minutes at 60°. The colour was checked and the dye simmered a bit longer to get more colour. Then it was strained.

I used alum-soaked yarn and added it to the dye. Although dock has natural tannins which act as a mordant, I prefer to use alum-mordanted yarn when trying a dye for the first time. You could try 2 skeins of yarn, one with alum mordant and one without, to see if there is any colour difference.

I worked on the ratio of 2:1 fresh leaves to yarn so I was able to dye 200g wool. The yarn was dyed in the usual way.

The roots of dock take a bit more preparation to make a dye bath.

First of all, I washed the roots to remove all the mud, then used a strong pair of scissors or secateurs to cut the roots into small chunks. The larger the roots originally, and the smaller they are chopped, the better the colour. Cut the roots into chunks of about 1 cm.

These root chunks are put into a large pot and covered with several litres of water. The pot is then boiled for several hours to release the colour. I found I needed to boil the pot again the next day until I had a good colour in the water. The dye was then strained, the mordanted yarn added, and simmered at 60°degrees for an hour before allowing to cool completely in the dye bath. Rinse and dry the yarn in the usual way.

What do you think of the results?

Although dock leaves can be used like vine leaves for Greek-inspired meze, I haven't tried it.

Dock root chunks

Dyed with dock, leaves on the left, roots on the right

It's been too hot to make a large project, so I've been crocheting hexagons to combine to make a cushion cover or blanket. I'm delighted with the way the natural colours from my dye experiments go so well together and remind me of honeycomb.

The pattern is super easy, here it is....

Crochet Hexagons

Crochet hexagons in natural honey colours

Using any yarn with the correct size hook- I used 4ply yarn and 3.5mm hook.
To Begin: Chain 4, insert hook into the first chain to make a round
Round 1 Ch3, tr, ch2 *2tr, ch* 5 times, slip st into 3rd chain of beginning st of the round. The 2ch spaces make the Hexi corners
Round 2 Ch3, tr into the top of st, tr into space, *ch3, tr into corner, tr into top of next 2 st, tr into same corner* 5 times, ch 2 tr into corner, sl st into 3rd ch of beginning st of the round.
Round 3 ch3, tr in next 2 st, tr into corner, ch 2, *tr into corner, tr into next 4 st, tr into corner, ch 2* 5 times, tr into corner, tr into top of st, slip st to 3 ch and cast off. Sew in the ends as you go.
Join the hexi shapes by holding them wrong sides together and Dcr along the edge to make a raised seam.

AUGUST

Poem for August

Cheerful summertime,
See the rich, bright blooms grow
In the perfect heat.

August is one of the hottest months of the year with average temperatures from 19° to 13° and only 9 days of rain. It is the month of summer holidays and enjoying being outside.

August was so named after the Roman emperor Augustus Caesar who had many important events in his life happening this month. The Anglo-Saxons called it 'Weod monath', not after mis-spelled woad, but meaning 'weed month' when weeds and plants grow most abundantly.

The first of August is Lammas, and was a day or period of thanksgiving for the harvest. On Lammas Day, farmers made loaves of bread and gave them to the local church, or they crumbled the stale bread and spread it in the corners of the barn to bring good luck for the harvest.

IN THE DYE GARDEN

August is always a busy month in the dye garden; watering, weeding, picking and drying the flowers. It's often a daily pleasure – wandering in the garden in the sun, picking the flowers that are in full bloom to use fresh in dye pots or to dry and use later. Currently I'm picking calendula (pot marigolds) coreopsis, dyer's chamomile, tansy, safflower and foraging for crocosmia (Montbretia).

This prolific, spreading bulb plant grows like a weed in the hedgerows of Cornwall in its bright orange form. The garden varieties come in a range of colours from pale citrus yellow to peach, orange and deep red. Interestingly they all give a bright yellow colour regardless of the original flower colour.

I have used them as a dye to give shades of yellow. It's not a long-lasting dye and does fade in sunlight over about 6 months, but it's very pretty and easy to pick and use. Simply pull the flowers from the stem, leaving the tiny bud flowers to grow for a second pick from the plant.

To harvest dye garden flowers, pick them when they are in full bloom, ideally in dry weather. Snap the flower heads of at the base of the flower or snip with scissors. Lay the fresh flowers on dehydrator trays in a thin layer and follow manufacturer's instructions for drying times etc. Make sure they are really dry before you store them because they will go mouldy and musty if they are still a bit damp.

Montbretia

TANSY (*TANACETUM VULGARE*)

Tansy has been used for centuries as a medicinal herb to treat intestinal worms, rheumatism, fevers and sores. In the Middle Ages, Christians began serving tansy with Lenten meals to commemorate the bitter herbs eaten by the Jews in the Old Testament on their exodus from Israel into the desert.

Tansy has a strong, pungent smell which has been used as an insect repellent for hundreds of years. It was also used as a rub for meats to repel flies and maggots. Research has shown it to be approximately 65% effective as an oil blend on the skin to repel insects. However, tansy can cause contact dermatitis if you have sensitive skin. To be safe, I always wear gloves when cutting and handling tansy. Tansy is quite tough so I use scissors to snip off the flower clusters, and then dry them like that. One of its other uses is as a companion plant: tansy planted with potatoes will deter the Colorado beetle.

Tansy as a dye plant gives gentle soft yellow colours.

AMARANTH AS A DYE

An unexpected gem in the dye garden this year is amaranth *(Amaranthus species)*. This stunningly coloured plant also gives a similarly bright dye colour.

I researched the method of extracting the dye colour – it's different to all the other dyes that I use because the colour is destroyed by heat. Therefore, it can't be simmered in a dye pot. It is also not very light-fast.

I discovered two methods to extract the colour. Both use a large jar, as follows:

JAR 1. WATER AND DAYLIGHT METHOD

Cut the seed heads, flowers, leaves and stems into small pieces and pack the jar to ¾ full. Top up with water, close the jar and leave in sunlight for a week. (I forgot mine and left if for two weeks andit didn't seem to matter.)

JAR 2. PH3 AND DARKNESS METHOD

Cut the seed heads, flowers, leaves and stems as before and pack into a jar.

Mix together water and vinegar to get a pH of 3.0 and pour this into the jar. Close the lid and leave in the dark for a week. (Or two weeks in my case.)

After a week (or two), strain the liquid from the jar though a fine sieve into a bucket and add alum mordanted wool. (Alum at 7% WOF) Leave overnight to absorb the colour.

Rinse gently in lukewarm water and allow to dry out of direct sunlight.

The 2 colours differed slightly:

The darker purple shade on the right was the water/daylight method and the bright pink shade next to it was the pH3/darkness method.

The acidity of the liquid in jar 2 has obviously brightened the colour which suggests that amaranth dye is pH sensitive like madder (March)

After the initial two test skeins were dyed, there was still colour left in both buckets so I mixed them together to exhaust dye. I used another skein of mordanted yarn – as before- and soaked the yarn in the dye bucket for 24 hours. I was able to do this 3 times with reducing tones of pink colour until the dye colour in the bucket was virtually gone. The colour will fade over time but is bright and beautiful while it lasts.

A BIT MORE ABOUT AMARANTH

Amaranth is an interesting plant and not just to dye with. It is easy to grow from seed and some species are short lived perennials. It is very showy in flower borders where it's structure and colour add a bright splash. In some parts of the world, it grows wild and self -seeds so prolifically it is known as 'Pig weed'.

The flowers are long and fluffy, a bit like catkins, which stay pinkish even after the plant has died. This gives one variety of the plant the name, 'Love lies Bleeding'.

Amaranth dye experiment

The leaves of some varieties of Amaranth can be eaten, and throughout the ages amaranth grain has been eaten by northern and southern American indigenous people. The Aztecs and Incas believed it has supernatural powers. Amaranth grain is gluten-free and is rich in protein and micronutrients. It has a wholesome, nutty taste on its own and can be quite a dense mixture (due to no gluten to give a dough elasticity.) Mixing with whole-wheat flour creates a lighter flavour and texture.

Amaranth in my garden

I have made both these recipes with great success:

Orange Amaranth Bread

- Ingredients
- 1 ¼ cups warm water
- 3 tbs honey or sugar
- 2 ¼ tsp active dry yeast
- 2 ½ cups bread flour
- ½ cup wholewheat bread flour
- ¾ cup amaranth flour
- 1 tbs wheat germ or wheat bran
- 3 tbs grated orange zest
- 2 tbs vegetable oil
- ¾ tsp salt

Method:

1. Stir water, honey and yeast together in a large bowl. Stand until the yeast softens and forms a foam.
2. Mix all the flours together.
3. Stir orange zest, vegetable oil and salt into the yeast mixture. Gradually add the flour to form a sturdy dough. Knead until smooth on a lightly floured surface, about 10 minutes.
4. Place the dough in a large, oiled bowl and turn it to coat in oil. Cover with a damp cloth and leave to rise in a warm place until double in size.
5. Grease a loaf pan.
6. Knead the dough again and form into a loaf shape. Place in the loaf pan and leave to rise until doubled in size again.
7. Pre-heat the oven to 350° F/170° C.
8. Bake the bread in the preheated oven until the top is brown and the bottom of the loaf sounds hollow when tapped.

Amaranth and Garlic cheese crackers

Ingredients

- 1 cup amaranth flour
- 3 tbs olive oil
- 3 tbs water
- ½ tsp baking powder
- 1 fat clove of garlic, finely chopped or grated
- ½ cup grated cheese
- ½ tsp salt

Method:

1. Pre-heat the oven to 355°F/ 180°C.
2. Place all the ingredients into a food processor and combine. Alternatively mix really well together in a bowl. The mixture will look like damp sand.
3. Remove from the bowl and knead gently on a piece of baking paper until smooth and no longer clumpy.
4. Flatten the dough and place another sheet of baking paper on the top. Roll out to about 5mm thick. Cut out rounds with a biscuit cutter.
5. Place on a baking tray and bake for 12-15 minutes. They will puff up a bit which makes them softer to eat. Bake until golden, then remove from the oven and cool on a wire cooling rack. Store in an airtight container.

Amaranth and garlic cheese crackers, serve with a generous chunk of cheese and crisp green apple.

SEPTEMBER

Poem for September

Still warm September,
Fresh blackberry picking
And rain watering

This month was the seventh month in the Roman calendar, and 'septum' means seventh in Latin. The Romans also believed this month was looked after by Vulcan, the god of Fire and forging so they would expect fires, thunderstorms and volcanic eruptions in this month.

The Anglo-Saxons called it 'Gerst monath', Barley month as this is when the barley was harvested to brew with. It was also called 'haerfest monath', Harvest month, and this is when we get the word harvest from.

The calendar changed from the Roman to the Gregorian one we use now, in September 1582. (Named after Pope Gregory XIII who introduced it and based on the astronomical year which is more accurate) In the changeover, 3 September instantly became 14 September and 11 days were 'lost'. There was protesting in the streets as people believed their lives would be shortened by 11 days!

Average temperatures range from 17° to 12° with an average of 10 days of rain.

IN THE DYE GARDEN

September is here! How has your dye garden been faring? I have been surprised at how much watering these plants needed throughout August but it has become a pleasant evening routine: watering the plants with rainwater from the water-butts, pottering, picking a few flowers each day. A lovely calm time at the end of the day.

I have been picking calendula, dyer's chamomile and woad leaves. The sunflower heads have ripened and the seeds can be used as a dye.

This month I have planted more dye seeds: Tansy, Woad, Weld and Soapwort (not a dye plant, but I thought I would give it a try as it might be a source of a natural soap to rinse yarns). They are now all in pots in the cold frame and hopefully will germinate and over-winter to plant in the spring.

ABOUT WOAD

Woad seedlings

Woad *(Isatis tinctorum)* is a flowering plant in the same family as cabbages and in fact its leaves are bit spinach-like. It has been used since the Stone age as a blue dye-giving plant and as a medicinal plant. It was certainly used by the Celts long before the famous Scot, William Wallace, AKA Braveheart, painted his face blue with it. It used to also be known as 'Asp of Jerusalem'. It grows fairly easily from seeds which should be from the previous year's flowers and look like tiny black, winged fruit. Older seeds tend not to germinate. It is sometimes grown for its yellow flowers but in some parts of the USA it is so rampant that it is seen as a weed. The bluest colour comes from plants that have been grown in rich soil with plenty of mulch and nutrients. Woad also likes a lot of sun.

The woad dye colour (Indigotin) is produced from fermented leaves, usually in a deep vat of water, alkali and a reducing agent The solution is whisked to create a froth which aerates the mix. The liquid is then strained and allowed to settle. The sediment at the bottom is the woad pigment. This can be dried to a powder or used straight away.

There are several ways to extract woad dye colour, which is a bit like indigo. Jenny Dean describes how to use fresh woad or Japanese indigo in her book, *Wild Colour*.

I was pleased to find a simple technique for woad. It doesn't give the strong rich blue that extracted woad gives, but does give a bright turquoise blue quickly and with minimal effort. You can check if your woad will give a good blue by sandwiching a leaf between two pieces of cotton fabric. Bash the leaf and it will release its leafy juice. This will initially be bright green but don't worry. As the colour oxidises in the air it will turn slowly blue.

I did some research and discovered a much simpler way to get a blue colour from the woad leaves I picked. Essentially woad is made into an ice/salt slushy! This is an adaptation of Rebecca Desnos' method.

DYEING WITH WOAD SLUSH

To prepare the woad slushy:
Place approx. 200g fresh woad leaves and 100g salt with a few ice cubes into an old blender and whizz to make a green mush.

Lay 100g skein of yarn on a sheet of clingfilm or waterproof covering on a flat, firm surface.

No need to mordant as woad is a substantive dye. However, as mentioned previously, when using a blue dye (normally used in a dye vat) use a mordant for better light and wash fast colours.

Wear rubber gloves (or have blue hands) and rub the slushy paste onto, and into, the yarn fibres. It will be a messy job and woad has quite an unpleasant cabbagy smell. A job to do outside perhaps!

Rub the slush onto the yarn for about 10 minutes then leave for a few hours until it stops turning blue. You should get a turquoise blue-green yarn with a lot less faff than setting up a dye vat.

Scoop/scrape as much of the paste off the yarn and save it. (See later)

Thoroughly rinse the yarn until the water runs clear and there are no bits left tangled in the fibres. This may take a while. Be careful to use tepid water and not to rub the yarn fibres because they may felt with rubbing and friction. Dry as normal.

With the left-over woad slush; put this in a pan and pour boiling water over it. Leave to cool until barely warm then add a skein of alum mordanted yarn. Simmer very gently at 60° degrees for about an hour then turn off the heat and leave to cool completely.

The skein should be a peachy pink. Such a beautiful contrast to the first woad colour.

2 dye colours from woad

SAFFLOWER *(CARTHAMUS TINCTORIUS)*

Another plant with the potential to give two colours is Safflower. It grows easily from seed and is like a small yellow thistle which flowers through the summer.

I picked the small stamens as they appeared and let them dry in a teacup on the windowsill. It's a bit like saffron and in fact, it's sometimes called 'Poor Man's Saffron' for the yellow colour the stamens give. However, Safflower can give two colours – yellow and coral pink.

DYEING WITH SAFFLOWER

The yellow dye comes first:

Put the dry stamens in a small gauze bag and soak in a bowl of cool water. Keep gently squeezing the bag to encourage the colour to flow. Once no more yellow colour comes from the stamens, this liquid can be used as a dye in the usual way.

Repeat the process with the 'used' stamens in a bowl of water with washing soda added to bring the pH up to about 11. The stamens will turn brown. Squeeze the gauze bag to extract the pinky red colour. Lift the bag out when no more colour runs and adjust the pH back down to 6 with white vinegar. Use this liquid as a second dye.

2 colours from Safflower – yellow from the first dye and peach from the second.

Note: I have since learned that the pinky red is on silk and cotton, not wool so the peach colour I got was lucky.

HOPI SUNFLOWERS

The Hopi people in North America have used sunflower seeds for centuries to dye cotton, wool raffia and baskets. The colours range from maroon, red, dark purple, blue and black.

I hoped for a strong grey colour but I was not very successful. The colour was disappointing. Perhaps I didn't use enough seeds to weight of fibre. I think I should try 2:1 next time.

If you fancy having a go with your Hopi sunflower seeds, here's what I researched and did:

Scrape the seeds from a large sunflower head onto a tray with a spoon.

Leave the seeds to dry out a bit then simmer in a pan of water for about an hour.

(Plenty of purplish black colour infused into the water so I felt hopeful about dyeing some yarn.)

Add 100g alum mordanted yarn into the cooled, strained dye.

Simmer at about 60° degrees for 30 minutes then leave to cool.

Although the yarn appeared to take up the colour from the water, it seemed to rinse away quite a lot. Perhaps I need a stronger dye solution next time, or perhaps the dye is simply fugitive and won't fix well under any conditions.

Chamomile Cake

Although the chamomile in the teabags is not the variety you can dye with, it is of the same family - Anthemis.
This recipe is a firm favourite in our family. It's definitely worth the effort and it makes a huge cake for parties and special occasions!

Ingredients:
- For the cake:
- 2 cups whole milk
- 6 chamomile tea bags
- 5 cups plain flour
- 3 tsp Baking Powder
- 1 tsp salt
- 3 cups white sugar
- 6 large eggs
- 2 tsp vanilla extract
- 1 ½ cups sunflower oil

For the Buttercream meringue:
- 4 large egg whites
- ¾ cup + 1tbs white sugar
- ½ cup honey
- 300g unsalted butter at room temperature
- 1 ½ tsp vanilla extract
- Pinch of salt

To make the cake layers:
1. Gently heat the milk and the tea bags in a small saucepan for about 5 minutes. Leave to cool and steep for about 2 hours.
2. Preheat the oven to 180°C and grease and line two 9-inch springform cake pans.
3. In a large bowl, mix together flour, baking powder and salt.
4. In another mixing bowl, add eggs, sugar and vanilla and beat well (ideally with an electric mixer or whisk) until the mixture is very thick and fluffy. Slowly drizzle in the oil, mixing well all the time, then beat for another minute.

5. Spoon by spoon, add the dry mixture to the eggs and sugar mix, and gently mix until just combined. Add the tea-infused milk and mix until smooth.
6. Pour the batter into the cake pans and bake for 40-50 minutes until golden brown, and a toothpick inserted into the centre of the cake comes out clean. Cool completely.

To make the buttercream meringue:

1. Place the egg whites, sugar and honey into a metal bowl over a pan of simmering water. Whisk constantly until it begins to thicken, then remove from the heat and whisk on high speed for about 7 minutes until the mix forms stiff peaks.
2. Slowly add the butter 2 tbs at a time and mix well before adding more. (The mix may look a bit curdled and soupy, but keep mixing!) Keep adding the butter and beat until the buttercream mix looks smooth. Add vanilla and salt. Place in the fridge to firm it up a bit if necessary.
3. Place one layer of cake on a serving plate, spread with half the buttercream mix, then put the other layer of cake on and repeat with the butter cream. Decorate as you wish.

OCTOBER

Poem for October

Windy October,
See a yellow leaf fall,
Deserting the bare twig.

In the Roman Calendar, October was the eighth month, 'Octo', meaning eight.

It is the season of misty mornings and bright cold afternoons picking blackberries and kicking up leaves. The leaves turn yellow, orange, red and brown as the chlorophyll (which makes the leaves green and helps plants photosynthesise) degrades and is replaced by other pigments which appear as autumn-leaf colours.

Average temperatures range from 12° to 9° with 14 days of rain.

IN THE DYE GARDEN

It is definitely becoming autumnal with gusty winds and leaves falling so the trees are almost bare already.

My dye-plant seeds, planted last month, have come up and are cosy in the cold frame. The dye garden has mostly been cut back and as most of the dye plants are perennials, they will easily over-winter in the mild southwest of Cornwall where I am. If you aren't sure about your dye garden over-wintering so well, try packing straw around the plants and covering them with hessian sacking.

I have dried the last of the flowers from the calendula in my dehydrator and they are now in jars ready to be used. Storing dried dye stuff is the easiest way to have an abundant choice of colours throughout the year. If you haven't got a dye garden, or haven't had much success this year, there are plenty of natural dye suppliers online. Search the internet, most companies will post mail order to you.

DYE SUPPLIES AND RHUBARB ROOT

It is very easy to obtain madder, rhubarb root, chamomile, cochineal, weld and logwood. All these colours give reliable results and are suitable for beginners and experienced dyers, and can be stored in jars until needed.

I have already used madder and logwood with you – rhubarb root is another one of my favourites. It gives mustard yellow and can be over-dyed for some beautiful colours and effects. Rhubarb root can be adjusted by pH or metal solutions to give a wider range of colours.

Here are a couple of ideas:

Try dyeing a skein or two of yarn with rhubarb root in the usual way, then twist it tightly back into a skein and put it into a madder dye bath. Where the yarn is tightly wound, the madder wont dye so well and you will have a tigerish effect.

Rhubarb root also dyes well in a solar dye jar, give it a go using the technique described in July. You can start solar dye jars at any time of year.

Try adjusting the base colour of rhubarb root dyed yarn like this:

Dye 5 mini skeins using the traditional dye pot method.

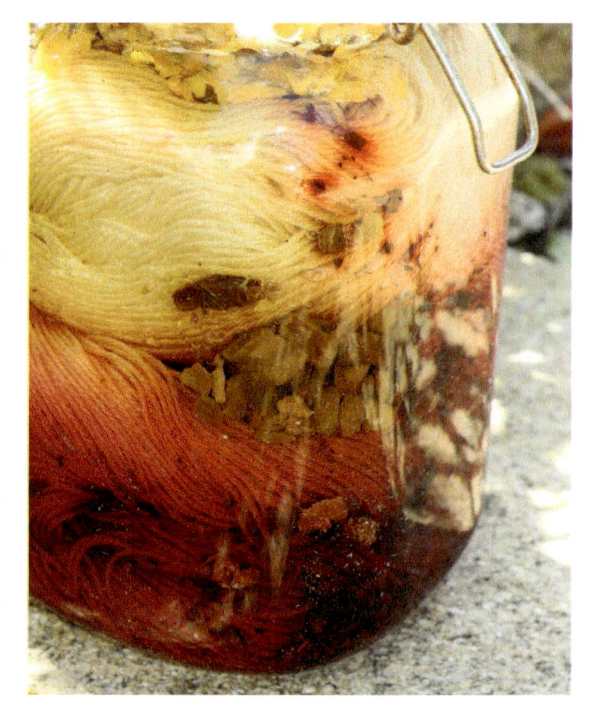

Once the skeins are dyed, keep one as the sample and use the other 4 as trial samples in the 4 adjustments.

Solar jar with rhubarb root

ACID PH

Pour about a quarter of the original dye into a smaller pot, add a bit more water if you need to. Then add white vinegar or lemon juice, a teaspoon at a time until you see the colour change. Stir well then add the dyed skein of yarn. Gently heat for about 10 minutes – the colour of the skein should change to a brighter yellow. If it doesn't change much, remove the skein, add a teaspoon or two more of vinegar/lemon juice and put the skein back in again.

ALKALI PH

Repeat the process as you did for acid (DO NOT heat), but substitute household ammonia or washing soda for the vinegar/lemon juice. Be VERY careful with ammonia, and avoid breathing the fumes or getting it on your skin.

IRON SOLUTION

Using the iron solution (as described in Terms and Techniques), add a teaspoonful at a time to a quarter of the dye solutions. The colour should go olive green or even grey/black. Add the skein and gently heat for 10 mins. Resist the temptation to add too much iron solution as it can make the yarn a bit crispy and brittle.

Rhubarb root samples adjusted with pH and metal solution modifiers

COPPER SOLUTION

Using the copper solution (as described in Terms and Techniques), modify the colour as for iron solution. The colour should be a bit green, almost a chartreuse green.

COCHINEAL

The bright pink colouring of Cochineal comes from a tiny scale beetle, *Dacylopius coccus*. The beetle feeds on the moisture of prickly pear cactus in central and south America and the Canary Islands. Cochineal colouring is a common additive to foods, cosmetics, medicines (as carmine red, E120) and as a natural dye. It has excellent light- and colour-fastness and was prized by dyers in the 1700s. It gives bright pink, fuchsia colours, reds and purples and exhausts well to give decreasing shades of pink.

When you buy cochineal natural dye, it may seem a bit more expensive than other dyes, especially when you see the tiny silver-grey bead-like beetles. However, those tiny insects contain a huge amount of colour potential and 25g can dye at least 500g of yarn. For this reason, Cochineal is a good choice to dye garment quantities of yarn.

USING COCHINEAL TO DYE WITH

The dry beetles need to be crushed to release most colour, and the simplest way to do this is in a bowl with a spoon, or in a plastic jug with the end of a rolling pin. Simply crush the majority of the beetles to miniscule bits, it doesn't need to be a powder.

Put the crushed beetles into a jug, (if not already in one) and pour over boiling water. The colour will instantly flood dark pink. Leave to cool then strain and pour this dark pink dye-liquid into your dye pan. Add plenty of water to the dye pan and dye your yarn in the usual way. If you use 25g cochineal to dye 500g yarn, there will be enough colour molecules in this dye liquid to dye to a fairly dark, rich pink.

There will still be more colour potential in the soaked and strained beetles. Add more boiling water and repeat the process. If you want to dye more than 500g of yarn, then use the first two or three colour extractions, using them all in the same dye pot at once.

This colour extraction process can be repeated four or five times with the initial 25g cochineal to get decreasing colour to a soft, baby pink. The cochineal dye is then said to be 'exhausted'. After giving all that colour, I'm not surprised its exhausted!

Here's a crochet pattern I have designed and made, using 500g of cochineal dyed DK yarn.

Loving the Lattice

By Caroline Bawn / Gorgeous Yarns

This simple lattice T-shape top was inspired by the beautiful lattice plasterwork ceiling at Lanhydrock house in Cornwall. It has clusters of double crochet diamonds connected with simple chains to give an overall lattice pattern which is repeated over 6 rows throughout. Crocheted with a 3.5mm hook, and seams finished with dc, it has 2 panels (front and back) and 2 panels for sleeves and a slash neck creating the simple T shape.

Pattern has 3 sizes, for 36"inch/92cm, (40"inch/102cm) and [44"inch/112cm] bust

Finished sizes are: 42"inch/107cm, (47"inch/120cm) and [52"inch/133cm] bust, length 24"inch/61cm

You will need...

Hand dyed DK ply 500g (600g) [600g]
3.5mm hook
Yarn needle to sew in ends

Special stitches & details

Lattice Panel
Row 1. 1ch, 1dc into each of next 2 dc, *3ch, 1dc into next 5ch arch, 3ch, skip 1dc, 1 dc into next 3dc, rep from * to end, omitting last dc at end of repeat, turn
Row 2. 1ch, 1dc into first dc, * 3ch, 1dc into next 3ch arch, 1dc into next dc, 1dc into next 3ch arch, 3ch, skip 1dc, 1dc into next dc, rep from * to end, turn
Row 3. 5ch, (count as 1tr and 2ch), 1dc into next 3 arch, 1dc into each of next 3dc, 1dc into next 3ch arch,, * 5ch, 1dc into next 3ch arch, 1dc into each of next 3dc, 1dc into next 3ch arch, rep from * to last dc, 2ch, 1tr into last dc, turn
Row 4. 1ch, 1 dc into first tr, 3ch, skip first dc, 1dc into each of next 3dc, *3ch, 1dc into next 5ch arch, 3ch, skip 1dc, 1dc into each of next 3dc, , rep from * to last 2ch arch, 3ch, 1dc into 3rd of 5ch at beginning of previous row, turn
Row 5. 1ch, 1dc into first dc, 1dc into 1st 3ch arch, 3ch, skip 1 dc, 1 dc into next dc, * 3ch, 1dc into next 3ch arch, 1dc into next dc, 1 dc into next 3ch arch, 3ch, skip 1dc, 1dc into next dc, rep from * to last 3ch arch, 3 ch, 1dc into 3ch arch, 1dc into last dc, turn
Row 6. 1ch, 1dc into each of first 2dc, * 1dc into next 3ch arch, 5ch, 1dc into next 5ch arch, 1dc into each of next 3dc, rep from * to end, omitting 1dc at end of row, turn

Foundation Treble Cast on
http://www.cherryheartcrochet.co.uk

Method

Cast on and work 137 (153) [169], turn, and work 1 row of dc each stitch, then
1ch, 1dc into each of next 3 dc, * 5ch, skip 3 dc, 1dc into each of next 5dc, * rep from * to * omitting second dc at end of last rep, turn
Then commencing with row 1 of the lattice panel, worked as set throughout, repeat the panel rows until work measures 24'/61cm, ending with a row 6, turn,
Work 1 row of 137 (153) [169] tr evenly across the row of dc and chain arches, turn
Work 1 row dc in each of the tr, cast off.

Sleeves. Make 2
Using foundation treble, cast on and work 129 stitches, turn, and dc each tr.
Then1ch, 1dc into each of next 3 dc, * 5ch, skip 3 dc, 1dc into each of next 5dc, * rep from * to * omitting second dc at end of last rep, turn
Working lattice panel as set, repeat the panel rows until the work measures between 5"- 7"/13 - 19cm, ending with row 6.

To make up
Gently steam block. The lattice has a bumpy texture which is pleasing and adds to the design, be careful not to squash this out of your work when you block.
Pin front and back pieces together, matching patterns at the sides and front to back. Dc across the shoulders from outer edge towards the neck each side for 4.5"/12cm, (6"/16cm) [7"/19cm]. (Leave the centre portion of the neck open.)
Fold sleeve pieces in half along the length of the row, mark the centre fold, pin this to the shoulder seam, and pin the sleeve in place front and back. Dc together.
DC from the edge of the sleeve, under the arm and down the side of the 'Loving the Lattice T' leaving a gap of 6"/16cm at the bottom edge to create a split if you wish and cast off. Repeat on the other side. Sew in all the yarn ends.

Abbreviations

alt – alternate
beg – beginning
bet – between
BL – back loop
bl – block
BPdc – back post double crochet
ch(s) – chain(s)
cl(s) – clusters
dc – double crochet
dec – decrease
dtr – double treble crochet
edc – extended double crochet
ehdc – extended half double crochet
esc – extended single crochet
fdc – chainless foundation double crochet
fhdc – chainless foundation half double crochet
FL – flont loop
FPdc – front post double crochet
fsc – chainless foundation single crochet
hdc – half double crochet
hk – hook

inc – increase
lp(s) - loops
p – picot
pat(s) – pattern(s)
pm – place marker
rem – remain(ing)
rep, *, [] – repeat(s)
rnd(s) – round(s)
sc – single crochet
sk – skip
sl st – slip stitch
sp(s) – space(s)
st(s) – stitch(es)
tog – together
tks – Tunisian knit stitch
tps – Tunisian purl stitch
tr – treble crochet
tr tr – triple treble crochet
tss – Tunisian simple stitch
ws – wrong side
yo – yarn over

Lastly, following a dark pink theme and being seasonal, here's a recipe you can try. (It's a much more successful way of using red cabbage than trying to dye with it!)

This is the season for mellow, comforting food like sausages and mash. One of the vegetable dishes we enjoy at this time of year is braised red cabbage.

Red Cabbage with apples

Ingredients

- 1 red cabbage, finely shredded
- 2 bay leaves
- 2 star anise
- 1/2 tsp cinnamon
- 100ml vegetable stock
- 50g brown sugar
- 75ml balsamic vinegar
- 2 apples, peeled, cored and finely chopped,
- 1 onion, finely shredded.
- A generous blob of butter

Method

1. Place all the ingredients except for the apples and butter in a pan. Bring to the boil then reduce heat to a slow simmer for about 30 mins, stirring occasionally. Add the apples and continue gently cooking for another 15 minutes.
2. Lastly add the generous blob of butter and stir through the cabbage.
3. Serve hot with sausages!

NOVEMBER

Poem for November

Darker November
A heavy, chilly frost creeps
Beneath the garden door

The name November comes from the Latin word 'novem' meaning nine as it was the ninth month of the Roman calendar. The Anglo-Saxons called it 'wind monath' for fairly obvious reasons.

The first of November is All Saint's Day. This used to be called 'All Hallows Day', (hallow was the old word for holy person) and it began with feasting the day before, the Eve of all hallows or hallows 'ene. The second of November is All Soul's Day when family members who have died are remembered.

On 5th November, Bonfire night is celebrated to remember Guy Fawkes and his gang who almost blew up the houses of Parliament in the Gunpowder Plot of 1605. The Catholic gang plotted to blow up the parliament and the protestant King James 1. The plot was discovered and the gang were captured, tortured and executed for treason.

Since 1918, 11th November has been commemorated as Armistice Day to remember fallen servicemen from any country in any conflict or war. Poppies are worn as a symbol of commemoration.

Temperatures typically range from 12° to 7° with around 15 days of rain.

IN THE DYE GARDEN

In the dye garden all the plants have been cut back to ground level and compost added as a mulch on top of the soil. I compost all the dye stuff and it gives me a sense of satisfaction to know that the natural cycle is complete: plant-dye-compost-plant

The seeds of woad, (Isatis tinctorum) weld (Ruseda leutola) and dyer's greenweed (Genista tinctorum) have all germinated and are cosy in the cold frame beside the shed. I'm keen to give woad another try next year after the results this year. The weld and chamomile plants in the tubs from this year may well overwinter and sprout again next year. Too much weld isn't a problem because it dyes such a lovely yellow and over-dyed with Saxon blue gives beautiful green shades.

Madder (Rubia tinctorum) is one of my all-time favourite dyes. It grows easily and spreads, so keeping it contained is a good idea. It has also been cut back and will pop up again in the spring with vibrant green shoots.

I have not found madder easy to grow from seed. It is slow and temperamental to germinate. I have had most success growing madder from root cuttings, either from a fellow dyer or online mail order from a natural dye plant specialist

Weld, Woad and Dyer's Greenweed

Madder needs to grow, spread and flourish for about 4 years before the roots have plenty of colour to dye with so it is wise to consider this before planting. Of course, you can dig it up and use the roots before the plant is 4 years old, but the colour dye will be less intense.

If you decide to grow madder and then need to remove it, be aware that even tiny roots will grow and spread again. Another reason for container growing perhaps....

MORE ABOUT MADDER

As a dried dye stuff madder is available on line and isn't too expensive to have a try with.

One of the reasons I like madder so much is the wide range of shades it gives; from dark brick red, through oranges and peach to pale apricot and pink. It exhausts well so 100g of madder can usually give 500g of shaded yarns as the dye gets weaker – giving an overall Ombre effect through the skeins.

Dyed with Madder

Madder is also moderately pH sensitive so shades can be shifted more yellow with acids like lemon juice or vinegar, or more red/pink shades with alkali like bicarbonate of soda.

Madder gives good colour in the dye pot and also in a solar jar. Its intense colour works well as a contrast to yellows like rhubarb root or chamomile.

Lastly, madder combines well with other dyes to give more red to the dye colour. Working with logwood, madder gives a richer violet purple shade rather than a more blue-purple.

Madder has traditionally been used to dye fibres and fabrics for hundreds of years. In 17th century France, madder was known as 'grand teint' meaning the colour it gives is reliable, consistent and lightfast. Certainly 'Turkey Red' was created with madder and was used in Turkish carpets until synthetic dyes were easily available in the late 1800's. William Henry Perkins' Mauvine was the first synthetic dye and it was discovered by accident in 1856.

When you buy natural dyes on line, consider the source of the plants if you can, where they were grown and, if possible, choose ethical producers. (Vegan dyers prefer not to use cochineal and Lac because they are farmed insects.) Many dye plants can be grown in the UK and sourced from reputable suppliers. Many other dye colours can be foraged throughout the year and dried for later use.

HOW TO STORE NATURAL DYES

Natural Dye storage

Cudbear lichen solution

I have a selection of clearly marked glass jars for all my dried dye stuff and mordants. I can easily and quickly see what I have and what supplies are running low. Fresh dyes like flowers that I grow in the summer are dried in the dehydrator and stored in jars.

The ratio of WOF to dye changes with dried dye stuff – usually its 100% WOF or less, whereas fresh can be up to 400%. (It's worth checking with the supplier or in a dyers' textbook with more formal information than here.)

Fresh dye stuff can be dried in paper bags in an airing cupboard if you don't have a dehydrator. If I have used a dye in a dye pot and I'm not ready to use it again, but there is still good colour potential in it, I pour the dye into a large lidded bucket or plastic container. Label the contents and the date and use within 6 months ideally. It may go a bit mouldy too. The colour can sometimes darken and dull a bit.

Alternatively, strain out the dye stuff and put it in a clear freezer bag, labelled, and pop it in the freezer. It's obviously essential to label it clearly. You don't want to defrost dye stuff thinking it's tonight's supper! I try to use frozen dye stuff up within 6 months. I haven't noticed any changes in colour but I haven't specifically tested before and after freezing colours against each other. I suspect colours may dull a little.

Some natural dye stuff needs infusion techniques to develop colours and these dye solutions can be stored for a lot longer. I was given some Cudbear Lichen (Orcholechia tarteria) which I infused and fermented in water and ammonia in a Kilner jar and then used the solution to dye with. However, I don't recommend any lichen collecting for use in dyeing. I was just lucky to be given this small amount by a lichen expert.

Indigo needs its own vat. This is created by one of several methods and provided the vat is maintained it will give colour for several days. I have found indigo very reluctant to dye protein fibres. It is much more effective with cellulose fibres like cotton and linen. The subject of indigo dyeing is vast – enough for another book!

November always seems to be wet, mild and cold here in Cornwall. We rarely get those lovely clear, bright blue, icy cold days – they are usually in January and February, so in November I'm often sat in the warm, knitting and planning for Christmas.

Knitting and planning for Christmas

I like this pattern very much. It's quick to make and suits men and women. Perfect at the neck under a coat in the winter weather.

Newgale Beach

By Caroline Bawn / Gorgeous Yarns

A snug wrap scarf/cowl with a plaited detail and buttons.

You will need...

1 x 100g skein any double Knit, measuring at least 225m
4mm knitting needles
7 buttons, yarn needles and scissors, 16 safety pins
Take approximately 1 m of yarn from the ball before you start, this is used to sew on the buttons.

Tension

Measured over the rib pattern, (unstretched) 38st x 29 rows = 4"inch/10cm Finished measurement 25.5"inch/65cm x 6"inch/15cm

Special stitches & details

Rib pattern
Row 1 P 1 (K2 P2) 15 times, K2 P1 (RS)
Row 2 K1 (P2 K2) 15 times, P2, K1 (WS)

Method

Cast on 64 stitches, using 4mm knitting needles and work 14 rows in rib.
Holding unused stitches on needle or stitch holder, with RS facing, work each strand (before plaiting) on 4 stitches:
Row 1 P1, K2, P1
Row 2, K1 P2, K1
Work these 2 rows a total of 28 times, ie each row 14 times each
Leave these 4 stitches on a safety pin, cut yarn, rejoin yarn on next 4 stitches and repeat.
Keep working these lattice strands across the row until you have 16 strands held on safety pins
Weave the strands in a plaited panel that pleases you, use the photo for reference. (Be careful to keep the strands flat when you weave them)
When you have the strands in a lattice you like, pick up and work the stitches from each safety pin as they lay in the row, using the rib pattern as set. Continue working on all these 64 stitches, in rib until your yarn has nearly run out. Cast off.
Fold the neck warmer around and cross the edges over so that a short edge lays on the end of a long edge.

Using the 1m of yarn you saved, (or another piece if you forgot that bit!) sew on the buttons on the second rib in from the edge, down the long edge at the opposite end of the lattice. See photo for detail. Use the gaps in the lattice as button holes, and don't sew them on too tight so the buttons 'sit' nicely in the lattice.

Abbreviations

K – knit
P – purl
st(s) – stitch(es)
m – metres
yds – yards
cm – centimetres
in – inches
mm – millimetres
g – grams
LH – left hand
RS – right side
WS – wrong side
Rep – repeat
MC – main colour
CC – contrast colour
PM – place marker
SM – slip marker

inc – increase as described in pattern
dec – decrease, usually by knitting of purling two together
Sl1 – slip one st (purlwise unless directed otherwise)
Yo – yarn over (also known as yarn forward or yarn round needle)
Kfb – knit into front and back of next stitch
K2tog – knit two together
K3tog – knit three together
K2togtbl – knit two together through back of loops
K1tbl – knit one through back of loop
cdd – centred double decrease – slip 2 sts knitwise together, knit 1 stitch, then pass 2 slip sts over the knit stitch
puk – pick up and knit into front of loop lying between stitch previously knitted and following stitch
sl1wyib – slip 1 with yarn in back
sk2p – sl1, k2tog, psso
Sl1 making ds – after turning the work, slip the first stitch purlwise with yarn in front. Then take the yarn over the needle, pulling the stitch from the previous row over the needle to form a double stitch. Work next stitch as instructed. When you reach a double stitch in a following row, simply treat as 1 stitch and knit the 2 strands together.

After you have been out, and come back in from the cold and wet, why not have 10 minutes peace with a cup of hot chocolate?

DECEMBER

Poem for December

Bleak December
A bright firelight dances
And warms the heart

December was the tenth month of the Roman Calendar and means 'tenth'. The Anglo-Saxons called it 'Winter Monath', or 'Yule Monath', both words we recognise today. The tradition of the Yule log is very ancient and has its origins in pagan rituals. Vikings celebrated Yule to honour the winter solstice (21st December) by selecting a huge oak-tree log from the forest and bringing it home to burn for many days. This celebrated long life, prosperity and light in the darkness. The yule log was often decorated with ribbons as it was brought from the forest. The Christmas tree as we know it, with baubles, candles, stars and the like is a German tradition popularised by Prince Albert and Queen Victoria.

December 21st is the day of shortest daylight hours, marking the winter solstice. This is the point when the north pole is at the furthest point from the sun on the earth's axis. In the most northerly countries of the world, the winter months see almost no daylight at all.

Christmas is the main celebration of December. Giving gifts is customary at Christmas and this may have originated in the Roman festival of Saturnalia on 17th December when gifts were exchanged and slaves were waited on by their masters.

Average temperatures range from 10° to 5° with 15 days of rain.

GIFTS

As it's coming up to Christmas and the holiday season, perhaps you have a creative friend whom you know who would enjoy natural dyeing too.

Many kits are available which range in price and contents. Some have the dyes and instructions, some include yarn or fibre etc. The kits may be for traditional method dyeing in a dye pot, or a solar dye method in a jar, even a tincture dye kit to create multi-coloured yarn in a steamer. If you have particularly enjoyed one method of dyeing, you could have a go at creating a gift yourself for your friend, perhaps include the yarn and your favourite pattern too. If that's too much, you could gift your creative friend one of your naturally hand-dyed skeins of yarn from this year or make something and give a hand-made present.

NATURAL DYES IN LIQUID FORM

A few natural dyes are available as 'tinctures' – concentrated liquids with the colour from the dye plant held in suspension. The tincture I use the most is indigo – Saxon Blue – which I buy from an online supplier and friend, Helen Melvin of Fiery Felts.

Indigo requires a vat set up and only dyes cellulose fibres successfully. So dyeing protein fibres any shade of blue requires traditional woad or Saxon Blue tincture.

It is bought as a liquid and used as per the supplied instructions to give a range of blues from dark teal and kingfisher blue, through mid-blue to pale icy blue depending on the concentration of colour to water in the dye pot. Saxon Blue usually exhausts completely and this makes it different from most natural dyes.

Other tinctures and dye concentrates are available as liquids or sometimes as powders that you mix with water. Think of these as super concentrates. Mix with water in a little bottle and use in a splash pattern for a multi coloured effect, a bit like confetti.

CONFETTI DYE EFFECT

To use these tinctures to create confetti yarn, pre-mordant your skein with 7% alum WOF.

Lay the damp yarn on a long sheet of greaseproof paper on a flat surface. Mix the tinctures with water in little bottles and literally drip, drop and splash onto the yarn in any way you like. You don't need to be artistic, the more colourful the better. Carefully wrap the yarn up in the greaseproof paper and roll into a spiral. Place this roll into a steamer and gently steam for 30 minutes. Allow to cool in the steamer before unwrapping.

Rinse the yarn gently then dry in the normal way. This yarn will be multi-coloured. Perfect for a special single skein project or as an accent with a solid colour.

WHAT ABOUT DYE COURSES?

These range greatly in price and subject. From a half-day beginner's course to a two- or three-day (or more) course on indigo dyeing or complex dye methods. Courses and workshops can be via the internet as a download, live via Zoom or in person at an organised location. The benefit of online is that you aren't limited by location, what fun to join a dye workshop on the other side of the world from the comfort of home and no jet lag! The disadvantage of course is that you don't get the pleasure of meeting others and joining in. Whichever you choose, enjoy the colourful fun, take a notebook and ask lots of questions.

There are many informal ways to expand your skills and knowledge too.
- Consider following your favourite natural dyers on social media and seeing what they do
- Watch YouTube and other sources online for demonstrations, information and new ideas from natural dyers around the world.
- Search out natural dyers at the yarn shows you visit and chat with them. They may run courses or have tips to share.
- Read magazines and books about natural dyes and dye techniques. The history of natural dyes goes back millennia and is certainly well documented from the 17th Century, all around the world. There may also be information in galleries and museums you visit.

THE LAST MONTH

The last month of the year and the last month of our journey together in this almanac of natural dyeing. I hope you have had fun and created some colourful dyes and dyed yarns this year. If not – I hope that I have inspired you to have a try one day. You can't really get it wrong, it's like making strong tea and soaking yarn in it! Most importantly, have a go and have fun.

Thank you for sharing this year with me.

All good wishes,

Caroline

RESOURCES

NATURAL DYE SUPPLIERS

Gorgeous Yarns www.gorgeousyarns.co.uk
Helen Melvin www.fieryfelts.co.uk
Wild Colours www.wildcolours.co.uk
George Weil https://www.georgeweil.com
Dyeing Crafts https://dyeing-crafts.co.uk

DYE SEED SUPPLIERS

Natures Rainbow https://www.naturesrainbow.co.uk
Wild Colours http://www.wildcolours.co.uk
Ria Burns https://www.riaburns.co.uk
7 Wells https://www.7wells.co.uk
Ditchling Museum https://www.ditchlingmuseumartcraft.org.uk

BOOKS ON MY SHELF

The Dyer's Handbook Dominic Cardon ISBN 978-1-78925-549-2
The Art and Science of Natural dyes Joy Boutrup and Catharine Ellis ISBN 978-0-7643-5633-9
The Modern Natural Dyer Kristine Vejar ISBN 978-1-61769-175-1
Journeys in Natural Dyeing Kristine Vejar and Adrienne Rodriguez ISBN 978-1-4197-4707-6
The Rainbow Beneath my Feet Arlene Rainis Bessette and Alan E Bessette ISBN 0-8156-0680-X
The Colours of Nature Anne Stovlbaek Kjaer and Louise Schelde Frederiksen ISBN 978-87-998678-1-3
Wild Colour Jenny Dean ISBN 978-0-82305-727-6